"A timely book that sets out the human dimension for successful mine closure. The author brings a unique perspective – daughter of a miner and indigenous background, alongside her professional corporate, public and private sector experience. These different aspects provide the basis for an informed analysis with concrete examples and lessons learned. This is a book for mine managers and government officials, not just community specialists."

—**Carolyn McCommon**, *Specialist in community engagement, resettlement, consultation, and social performance in the oil, gas and mining sectors (Retired Global Community Advisor, Rio Tinto; consultant to USAID, IFC, UNDP and the World Bank)*

"A wonderful book. Avoiding the sterility of theoretical abstraction and universalist environmental presumption, it looks closely at the human issues of mine closure in very real, dramatic circumstances through the eyes of an author uniquely qualified to see the local and indigenous texture."

—**Bruce Harvey**, *Director, Resolution88 (Social License Solutions) and Adjunct Professor, Sustainable Minerals Institute, The University of Queensland, Australia*

"Mining companies enter a social terrain with inordinate amounts of political power, state-support and capital, leading to a complete reworking of not only the environment, but also the social fabric. Yet, the mines have limited lifespan, and once the mineral resources are exhausted, the operations must close. What happens to the local community then? Until today, mine closure, and its impacts on not only the economic life of the region but the cultural and spiritual practices of everyday life, are poorly known. Social Terrains of Mine Closure in the Philippines makes an important contribution to one of the little debated areas in the growing literature on mining-community relationship by showing not only the widespread disruptions that occur, but also by indicating ways how these disruptions could possibly be addressed. The case studies the author presents will be highly valued for future researchers and practitioners in the field."

—**Kuntala Lahiri-Dutt**, *Senior Fellow, Resource Environment & Development Program, Crawford School of Public Policy, ANU College of Asia and the Pacific, The Australian National University*

"This pioneering book gives authoritative and heart-felt insights into the conflicts between the exploitation of natural resources in developing countries and issues faced by indigenous communities, their future social well-being and post-mining landscapes which now underwrite mine closure policy, planning and management for both existing and potential or new mining ventures. Based on historic and ongoing case studies in the Philippines, Dr Chaloping-March offers to the metalliferous mining industry at large the much-needed understanding and sensitivity towards the interlacing concerns about mining and mine closure in developing countries."

—**Alan Baker**, *Centre for Mined Land Rehabilitation, Sustainable Minerals Institute, The University of Queensland, Australia*

"An original erudite ethnographic work offering an intuitive lens into composite factors including legislation, local politics, and community claims for entitlements. A distinctively local and indigenous viewpoint on intricate human-environment and governance issues surrounding mining. A must read for minerals industry professionals, and others engaged in resource development and environmental management including academics, NGOs and policy makers."

—**Velma I Grover**, *Adjunct Professor, Faculty of Environmental Studies, York University, Canada*

Social Terrains of Mine Closure in the Philippines

The current discourse on mine closure is informed predominantly by industry and corporate perspectives and predicated by experiences of mainly mining companies that are based in developed countries where necessary planning frameworks and regulatory requirements are well-established. Mine closure planning, well promoted and accepted as good business practice in the global minerals industry, has been primarily technical and precautionary both in approach and focus. Planning, modelling, and monitoring strategies incorporate comprehensive and detailed elements such as properties inherent in landforms, climate, geology, flora and fauna, among others. However, locality-based concerns that revolve around resource access and tenure, rights and entitlements tied to locality and indigeneity, labour recruitment, and other non-bio-physical elements are hardly examined. Any mine closure program that omits these elements is deficient and therefore ineffective.

Social Terrains of Mine Closure in the Philippines, based on ethnographic research and archival materials, presents the varying experiences of three mines to demonstrate that the mine closure process is an intense locus for competition and compromises among various social actors.

This book offers key messages for understanding the complex sociocultural, economic, political, and business realities that make up the social terrains of mine closure, and will be of great interest to students and researchers in development studies, community development, business studies, anthropology, and sociology. It will also appeal to those working in the global minerals sectors and NGOs that engage in development work and advocacy for responsible mining.

Minerva Chaloping-March, who grew up in a mining district in northern Philippines, combines consulting and research on social dimensions of mining, cultural criteria for mine closure, and transitions of mining communities. She is currently an Honorary Research Fellow at the Philippines – Australia Studies Centre, La Trobe University, Melbourne, Australia.

Routledge Studies of the Extractive Industries and Sustainable Development

African Artisanal Mining from the Inside Out
Access, norms and power in Congo's gold sector
Sarah Geenan

Mountain Movers
Mining, sustainability and the agents of change
Daniel M. Franks

Responsible Mining
Key concepts for industry integrity
Sara Bice

Mining in Latin America
Critical approaches to the new extraction
Edited by Kalowatie Deonandan and Michael L. Dougherty

Industrialising Rural India
Land, policy, resistance
Edited by Kenneth Bo Nielsen and Patrik Oskarsson

Governance in the Extractive Industries
Power, cultural politics and regulation
Edited by Lori Leonard and Siba Grovogui

Social Terrains of Mine Closure in the Philippines
Minerva Chaloping-March

www.routledge.com/series/REISD

Social Terrains of Mine Closure in the Philippines

Minerva Chaloping-March

Routledge
Taylor & Francis Group
LONDON AND NEW YORK
earthscan
from Routledge

First published 2018
by Routledge
2 Park Square, Milton Park, Abingdon, Oxon OX14 4RN

and by Routledge
711 Third Avenue, New York, NY 10017

Routledge is an imprint of the Taylor & Francis Group, an informa business

British Library Cataloguing-in-Publication Data
A catalogue record for this book is available from the British Library

Library of Congress Cataloging-in-Publication Data
A catalog record for this title has been requested

ISBN: 978-1-138-03877-6 (hbk)
ISBN: 978-1-315-17718-2 (ebk)

Typeset in Times New Roman
by Apex CoVantage, LLC

Contents

	List of figures	viii
	List of images	ix
	List of abbreviations	x
	Acknowledgements	xii
1	Mine closure: anticipating the end before the beginning	1
2	The socio-political landscape of mining in the Philippines	17
3	Pre-closure challenges at the Padcal mine	40
4	Company relations with communities around the Padcal mine	63
5	Antamok: launchpad of industrial mining in the Philippines	80
6	Change and continuity in Acupan-Balatoc	100
7	Drawing lessons from Itogon and Tuba for mine closure challenges	120
	Index	133

Figures

1.1 Varying domains of social actors in a mine closure social terrain 4
3.1 The tenement area of Philex Mining Corporation and its minesite camp in Padcal showing location within two barangays of two adjoining municipalities 47
5.1 Antamok, Acupan, and Balatoc as part of a large mining niche in Itogon, Benguet 82

Images

3.1 Partial view of Philex, the minesite camp in Padcal, Tuba, Benguet 45
6.1 One of the remaining bunkhouses in Balatoc 102

Abbreviations

ACMP	Acupan Contract Mining Project
ACUCCLA	Acupan Upper Camps Community Livelihood Association, Inc.
AGDIIS	Australian Government Department of Industry Innovation and Science
AGP	Antamok Gold Project
ARTC	Area Research Training Centre
BAGO	Baguio Antamok Gold Project
BC	Benguet Corporation
BCI	Benguet Consolidated Incorporated
BCMC	Benguet Consolidated Mining Company
BMTV	Benguet Mining Tourism Village
BWD	Baguio Water District
CAR	Cordillera Administrative Region
CBCP-NASSA	Catholic Bishops Conference of the Philippines – National Secretariat for Social Action Justice and Peace
CEC	Center for Environmental Concerns – Philippines
COMP	Chamber of Mines of the Philippines
CPA	Cordillera Peoples Alliance
CRD	Community Relations Division
DAO	DENR Administrative Order
DENR	Department of Environment and Natural Resources
ECC	Environmental Compliance Certificate
EGF	Environmental Guarantee Fund
EIS	Environmental Impact Statement
EITI	Extractive Industries Initiative
EO	Executive Order
ESL	Engineered sanitary landfill
FMRDP	Final Mine Rehabilitation and Decommissioning Plan
FPIC	Free Prior Informed Consent

FTAA	Financial and Technical Assistance Agreement
ICCs/IPs	Indigenous Cultural Communities/Indigenous Peoples
IPO-APSSOL	Indigenous Peoples Organisation – Alang, Pokis, Sabiang, Sta. Fe, Oliba, and Loacan
IPRA	Indigenous Peoples Rights Act
LCMCo	Lepanto Consolidated Mining Company
LIPMA	Loacan Itogon Pocket Miners Association
LRC-KsK	Legal Rights Center – *Kasama sa Kalikasan* ('Friends of the Earth')
MGB	Mines and Geosciences Bureau
MGDS	Mines and Geosciences Development Services
MISN	Mining Injustice Solidarity Network
MMSD	Mines Minerals and Sustainable Development
MMT	Multi-partite Monitoring Team
MOA	Memorandum of Agreement
MPSA	Mineral Production Sharing Agreement
MRFC	Mine Rehabilitation Fund Committee
NCIP	National Commission on Indigenous Peoples
NGO	Non-government organisation
NSCB	National Statistics Coordination Board
PD	Presidential Decree
Ph-EITI	Philippines – Extractive Industries Transparency Initiative
PMC	Philex Mining Corporation
PMRB	Provincial Mining Regulatory Board
PO	Peoples Organisation
PPO	Philex Purok Organisation
PSEC	Philippine Security Exchange Commission
RA	Republic Act
RFFC	Rapu-rapu Fact Finding Commission
RPP	Rapu-rapu Polymetallic Project
SDMP	Social Development Management Plan
TIHCAPOSI	The Indigenous Host Communities Association of Philex Outlying Sitios Inc.
TNMZ	Temporary No Mining Zone
UHA	United Highland Association
UNEP	United Nations Environment Program
VPO	Vice President for Operations
WB-IFC	World Bank – International Finance Corporation

Acknowledgements

I am profoundly indebted to my husband Peter March and daughter Nathania who have been most understanding, particularly when I could not join them for more exciting activities, due to writing tasks. I am incredibly grateful to my brother Wilbur Chaloping for preparing the diagram and maps for this book. To the rest of my family, thank you for the ever-available moral support.

This book is based on four (out of 11) chapters of the dissertation produced from my PhD study that was funded by the Australian Postgraduate Award for which I am tremendously thankful. I cannot name all the 328 men and women, most of whose names appear in my an appendix of my dissertation, and scores of more names in the volumes of my hand-written fieldnotes. I thank them all for sitting down with me during one-on-one conversations, paired interviews, and focus group discussions. They include wage-earning mine workers (some current and some retired), small-scale miners, company managers, corporate executives, key officers of government agencies, local government officials (barangay, municipality, provincial levels), community leaders, and representatives of NGOs.

I began fieldwork in Tuba and Itogon in July to September 2003, then continued in October 2004 to May 2005. Afterwards, my twice-yearly family and research trips to my homeland, the Cordilleras in northern Philippines, from 2010 to 2016, have kept me abreast of the issues I examined. It has been enriching going back many times over to my research sites in Tuba and Itogon, which has been easy to do due to their proximity to Baguio. When the prospect for publishing this book became certain in late 2016, my last visit to Baguio-Benguet in February 2017 included a few final interviews with barangay officials as well as a few pocket miners in both Loacan and Virac.

My deep thanks to the farmers, 'pocket' gold miners, and elders who received me in Nay-en, Sais, Sinko, Palunupon, Paday, Torre, Balayan, Kinmabab, Soyocto, Sal-angan, Santa Fe, Banget and other sitios of Ampucao and Tuba. I cannot thank them enough for the information they trustfully shared, their accounts of everyday living, challenges, and aspirations for the future particularly in relation to mine closure. Their wholesome wit and warm humour made every visit to their homes, farms, mining sites, and neighbourhoods a congenial learning experience for me.

At Philex Mining Corporation, my sincere gratitude goes to the late Engr. Leonardo Josef, and Mrs. Redempta Baluda. I owe special thanks to Diana Alegria, Jun Diso, and Badong Bannawi for accommodating me as an informal team member at their community relations office and tagging me along to the medical missions in the sitios around Padcal as well as in the inaugural workshop for the company's social development management program. Many thanks to the industrious and cheerful staff of the guest house in Philex for the evenings of endless conversations and pleasant laughter. My heartfelt appreciation to the many mine workers for sharing their enriching life histories. My intense gratitude goes to Janice Ommas Terso for her friendship and assistance in the research, and Atty Eduardo Aratas for devoting time to read and comment on Chapters 3 and 4 of the book.

In Acupan, my cousin Mr. Camait Cawaren and his family graciously provided me a shelter where to hang around, several sumptuous meals, and guidance for my going around Campo Singko, Campo Tres, Camp 1000, Camp Lolita, Camp 11, and Tang-tangadan. In Balatoc, former Virac Barangay Kapitan David Ecwasen warmly welcomed me, escorted me on the first few days and introduced me to more participants in the research. My appreciations also go to the current Barangay Kapitan Noel Bilibli for being receptive for conversations even without me securing any appointment. At the Acupan Contract Mining Project site, my gratitude goes to Engr. Angelito Gomez, Engr. Jacob Melecio, Engr. Ernesto Manipon, Mr. Gilbert Tanding, and Mr. Edward Kilakil, who willingly allowed their work to be interrupted to enrich me with their invaluable knowledge.

In Antamok, the presence of my childhood friends (back in Lepanto, Mankayan) Beverly Madanes and Nelson Maglanga gave me ease and safety to go around Surong and Luneta. I am thankful also to Barangay Kapitan Henry Diamnuan for the opportunity to learn a lot about Loacan particularly the place's history and peoples, livelihood challenges, and rituals involved in gold mining.

In Manila, I am grateful to the officers of the Chamber of Mines of the Philippines for their generosity in sharing their invaluable insights about the minerals industry and its challenges: Mr. Gerry Brimo, Engr. Art Disini, and Mrs. Nelia Halcon. Similarly, senior officers of mining companies kindly accommodated me in their offices for long conversations: Mr. Brian Yap, Mr. Marcelo Bolaño (now retired), and Mr. Tony Leviste. I am also indebted to Maria Mignon de Leon for reviewing Chapters 5 and 6 and providing comments. In Quezon City, I am tremendously indebted to Director (now retired) Engr. Horacio C. Ramos of the Mines and Geosciences Bureau for sharing his extensive knowledge and elucidating views of the sundry and difficult issues pertaining to mining. I also thank Mr. Reynaldo Alcanses of the Environmental Management Bureau for being equally considerate to accommodate my many questions.

I wish to thank my friend Dr. Augustine Doronila from whom I have continued to learn about the natural science side of mine closure. My supervisor Associate Professor Martha Macintyre encouraged me early on to publish my first academic book. I am exceedingly grateful to her for believing I can do it. Her trust is humbling yet challenging. Associate Professor Glen Banks and Professor Marcello Veiga unwittingly, through their examiners' reports, set alight my interest to publish. I intensely appreciate the encouragement and inspiration I took from their reports.

My highest praise belongs to the Almighty God whose gracious blessings include the daily strength and courage I needed during my fieldwork, and the vital sanity throughout the rest of the journey. I dedicate this book to the treasured memory of my father, a wage-earning underground miner in Lepanto, Mankayan. His life of dedication to work, and leadership in the community, are my pride.

1 Mine closure

Anticipating the end before the beginning

> Contractors and permittees shall technically and biologically rehabilitate the excavated, mined out, tailings covered and disturbed areas to the condition of environmental safety, as may be provided in the implementing rules and regulations of this Act. A mine rehabilitation fund shall be created, based on the contractor's approved work program, and shall be deposited as a trust fund in a government depository bank and used for physical and social rehabilitation of areas and communities affected by mining activities and for research on the social, technical and preventive aspects of rehabilitation. Failure to fulfil the above obligation shall mean immediate suspension or closure of the mining activities of the contractor/permittee.
>
> Chapter XI, Section 71, Philippine Mining Act of 1995

Introduction

As an economic activity, mining is centuries-old. However, planning to close a mine, and implementing a mine closure plan, are relatively recent practices in the minerals industry, having begun in the mid-1970s (Mulvey, Baker, & Scott, 2012). Their recent adoption is part of the mining sector's response to the public's growing consciousness of human health and safety issues including enduring criticisms about the repugnant impacts of abandoned mines. Beginning at least in the 1990s, mining has been 'losing its status as both a preferred land use and preferred economic activity', and has faced an increasingly stringent regulatory framework, at national and international levels, directly and indirectly (Pring, Otto, & Naito, 1999, p. 41). As part of responding to pressure from society at large, the international minerals sector has positioned itself to relate its operations to the values that will enable it to become viable.[1] Planning for closure has become an essential part of the global minerals industry's policies and approaches in line with objectives of progressive environmental and social performance. In the minerals sector, it has become almost an edict that mine closure planning must start from mine inception and the plan must be reviewed and updated throughout the entire life-of-mine. The aim is 'to ensure, as far as practicable, that the post-mining landscape is environmentally, socially and economically sustainable' (Mulvey et al., 2012, p. 2).

Currently, the discourse on mine closure is informed predominantly by industry and corporate perspectives and predicated by experiences of mainly mining companies that are based in developed countries where necessary planning frameworks and regulatory requirements are well-established. Mine closure planning, well promoted and accepted as good business practice in the global minerals industry, has been primarily technical and precautionary both in approach and focus. Planning, modelling, and monitoring strategies in mine closure incorporate as comprehensive and detailed as possible elements such as properties inherent in landforms, climate, geology, flora and fauna, among others. However, locality-based concerns that revolve around resource access and tenure, rights and entitlements tied to locality and indigeneity, labour recruitment, workers' mobility, and other non-bio-physical elements are hardly examined. Any mine closure program that disregards these elements is deficient and therefore ineffective.

The current approach of mine closure planning is highly technical-managerial involving stage-by-stage requirements which are mostly guided by legislation and principles of risk prevention and litigation avoidance. The factors shaping the design, implementation, and monitoring of mine closure strategies are international in dimension. This is a positive development because global mining companies leverage themselves against each other in developing stringent standards of performance for themselves. However, the focus has been on new and future mines which neatly fit the formulations of a 'planning-before-commencing' approach that assumes a blank-slate as a starting point for the mine closure process, envisioning a post-mining future.

Aim and approach

The book brings attention to the historical, political, and socio-cultural elements that impinge on the overall nature, tempo, and form of initiatives associated with mine closure and relevant activities such as rehabilitation and decommissioning. Convoluted issues concerning historical claims to land, sense of rootedness in homelands, and continuity of economic activities – which all have direct or indirect connections to a mine operation in a locality – do emerge or re-emerge and are stirred, during mine closure. These factors configure the types of tasks, relationships, timing and momentum, and resources for mine closure. Thus, the mine closure process is a social episode in the ebb and flow of life within communities. At the same time, the mine closure process becomes a strategic locus of interlacing and contradicting interests, values, objectives, and agendas of diverse social actors who all attempt to influence – in varying but allowable extents possible – the hazy course towards a post-mining future, however nebulous it may be.

This book presents three mines which are decades-old. They are not new or about to commence, but they offer important insights and tangible closure scenarios. They provide valuable cues to the twenty-first century mining companies about operational and closure realities in developing countries and remote areas which, incidentally, continue to attract global exploration and mining projects. Two companies operate the three mines under consideration: the 112-year old Benguet Corporation (BC) and the 60-year-old Philex Mining Corporation (PMC). BC closed its Benguet Gold Operations (BGO) in the early 1990s but after almost a decade, reopened the Acupan mine, one of BGO's several mines, and has since been operating until the present. At the same time, however, the company has been attending to rehabilitation concerns of its non-operating open pit in Antamok, a separate mine that closed in 1998. The company endeavours to address regulatory requirements while it deals with social and ethical issues. Meanwhile, PMC's Padcal mine is currently moving towards closure by 2022 which would be more than 65 years since operations commenced. The company began working on a mine transition plan in 2006, and the details of a post-closure future are yet to be substantiated and clarified in coordination with local governments and surrounding communities.

The two companies commenced operations at an era when the doctrine about mine closure planning was virtually unknown in the mining lexicon – more than six decades to over a century ago. However, the Mining Act, which was enacted in 1995 but whose constitutionality was established after a decade, mandates the mine operators to implement today's standards of managing mines including closure, rehabilitation, and decommissioning.

Applying an actor-oriented analysis (Long, 2001, 2004), mining companies are key actors in a social terrain that involves several communities that have evolved with mining operations. Taking leads from the mandates of the law, the mining companies are navigating, sketching, and actualising their trajectories which are sometimes clear, although oftentimes, their own goal posts shift. Figure 1.1 illustrates that global realities (e.g., trading systems, metal prices, debates about sustainable development and mining, etc.) directly and indirectly influence a country's national goals and economic priorities. In turn, national policies define resource and environmental governance. However, it is the local domain (Ovals A and B) that largely configures the timing, tempo, and nature of mine closure processes. Within Oval B are factors such as, for example, history of mining in the locality, livelihood activities and economic traditions, workers' mobility and labour

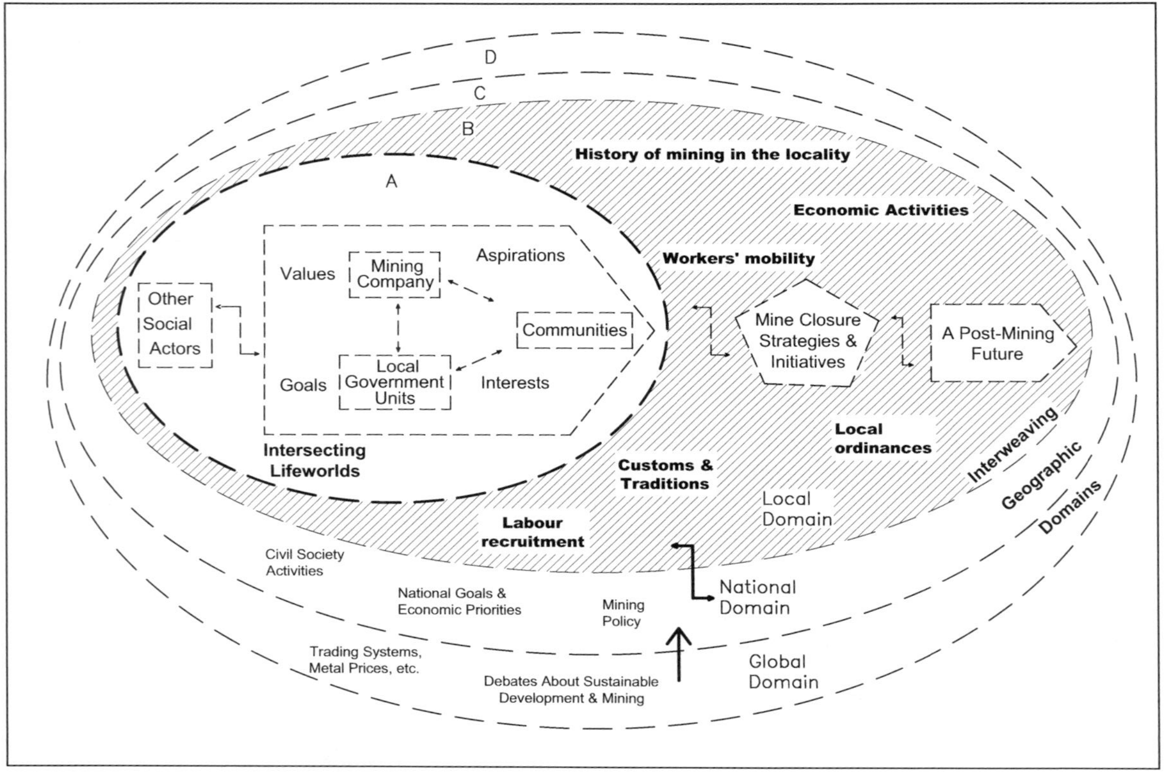

Figure 1.1 Varying domains of social actors in a mine closure social terrain

recruitment, among others. They make up the 'micro-scale' contexts and localised niches which are connected to wider 'macro-scale' phenomena (Long, 2003, p. 49). Across these scales are intertwining relationships and actions that intersect various social, symbolic, and geographic domains (represented by layered ovals). The domains are interweaving because they have permeable 'boundaries' (demarcated by broken lines).

The idea of 'battlefields of knowledge' (Long & Long, 1992) is useful in understanding intersecting lifeworlds of social actors. Perspectives, motivations, and values are heaved against each other as mining companies, local governments, communities, civil society entities, small-scale miners, policy makers, and others make claims and counterclaims to pursue their respective interests. The social terrain becomes a locus for some actors to legitimise and perpetuate their position of power, for others to find meaning to their rhetoric, and still others to justify their resistance against the current state of affairs.

To a mining company, a mine closure undertaking is a project that it implements in compliance with closure regulations. To communities, the mine closure process which involves their participation in many forms and at varying degrees, both voluntary and involuntary, is a social episode in their pathways and transitions. Implementing a mine closure plan is a form of an intervention which, in development theory and practice, aims to bring a positive change. In this case, the hoped-for change is a post-mining future towards which mine closure is projected. An intervention is more than 'projects of doing' (van Ufford, Kumar Giri, & Mosse, 2003, p. 31). This means that a mine closure undertaking goes beyond fixation to minutiae 'to-dos' and 'how-to' techniques beyond the protocols of engineering, financing, and complex technical elements involved in planning, rehabilitation, and post-closure monitoring. Rather, it also involves bridging, negotiating, accommodating, and contesting among actors. Interactions involve 'aspects of power, authority and legitimation' (Long, 2003, p. 55). Characterised by human agency, social actors engage in 'creating room for manoeuvre' to influence the material and social worlds to tilt outcomes of processes towards their favour. Creating this 'room for manoeuvre' involves granting consent, making negotiations, and haggling and bargaining towards reaching concessions, all of which apply some level of power. Power engenders resistance but also stirs strategic acquiescence (Long, 2003, p. 54). As power is often characterised by imbalance, mine closure initiatives are more likely to generate, trigger, and aggravate social disparities and conflicts between social groups than bring about mutual perspectives and shared values (Long, 2004, p. 31).

As used in the book, the term 'mine closure' refers to the discontinuation of mine operations, usually marking the end of employment of massive numbers of mine workers. It encompasses *decommissioning*, which involves removing mining-milling structures and other associated facilities and installations, as well as *rehabilitation*, which incorporates a series of activities to address safety and health concerns and strategies towards creating a new use for the former mine land. In some instances, the book uses the terms 'mine closure project' and 'mine closure undertaking' to refer to the set of initiatives required in decommissioning and rehabilitation. Another key term in this book is community, which refers broadly to groups of people (residents, villagers, townsfolk) occupying a territory. I also use the term 'minesite camp' which I interchange with 'occupational community' to refer to the socio-spatial settlements of mine workers and their families. They are located within the mine area as part of a mining company's scheme to ensure a stable and readily available workforce for productive activities. In referring to other similar socio-spatial units, such as those around the mines, I use the local terms *barangay* ('village') and *sitio* (a sub-village, component of a barangay).

By 'civil society', the term refers to 'the middle ground' of social relations between the socio-economic structure and the state' and includes 'the broad field of private, formal voluntary association – including the variety of social, cultural, and community organizations, churches, labor unions, neighbourhood associations, advocacy groups, and others' (Rhomberg, 2004, p. 9). Civil society activities support 'the emergence of various public spaces in which people learn about, discuss, and form opinions on social and political concerns independently of the state and the formal institutions of the public sphere such as mass media, legislatures, and courts' (Rhomberg, 2004, p. 9).

An inevitable end

Every mining operation eventually terminates when the minerals being extracted get depleted or when operating is no longer profitable. When the cost of producing the commodity – whether gold, silver, copper, nickel, or others – outstrips the return on selling it, the mine operates at a deficit. If this situation continues for a considerably protracted period, the mine will close business even if it still has vast reserves of minerals. Every mine operation has an anticipated closure date based on the estimated life of the mine, among other factors. Depending on the circumstances constituting a mining operation's shutting down, the mine may possibly reopen and resume production in the future because not every mine's closure is due to resource depletion (Laurence, 2006). In other instances, socio-political factors such as community unrest, labour disputes, peace and order problems, change in administration, or civil war cause some mines to close. Similarly, not every projected date for closure will happen because changes may take place in technology, costs, policies, governance regimes, and social situations.

The handbook *Mine Closure* which forms part of the comprehensive series *Leading Practice Sustainable Development Program* (LPSDP) publications (AGDIIS, 2016), underscores a tenet: the endpoint of mining is 'mine completion'. Mine completion is that state of affairs when the former mine land is relinquished to the next user who ascends to the responsibility of using it, and the mining company walks away. Reaching this endpoint requires final decommissioning of the operations site and rehabilitation of the area. Mine closure 'culminates in tenement relinquishment' following sign-off of responsibility (Peck et al., 2005, p. 4).

Mine closure: the literature, ideals, and realities

The literature on mine closure describes approaches and techniques in designing, planning, financing, implementing, and monitoring closure and post-closure processes. There has been an impressive growth of literature on mine closure and its associated themes which are produced by individual mining companies, government regulatory bodies, and industry associations led by the London-based International Council on Metals and Mining (ICMM). An ample contribution are the numerous papers presented in conferences such as the annual International Mine Closure Conference that will be in its twelfth year in 2017. The existing body of knowledge comes in numerous forms: handbooks, monographs, conference papers, journal articles, project reports plus manuals or guidelines by industry- and government-initiatives. They include company reports on mine closures implemented in the context of applying corporate standards and promoting corporate 'good practice' as part of self-promotion or building corporate image. International organisations such as the World Bank (WB-IFC, 2002; World Bank Multistakeholder Initiative, 2010), the United Nations Environment Program (UNEP)[2] have also produced both brief and exhaustive publications all of which add to an extensive body of knowledge.

The literature is exceptionally wide-ranging. According to the focus of the investigation, there can be at least four identifiable, often overlapping, strands of scholarship: a) environment, land and water; b) planning as part of the life of mine cycle; c) support systems for mine closure; and d) social aspects. A quick run-through of themes, not a detailed dissection, would suffice here to present a backdrop of knowledge gaps which this book can help fill.

Environment, land, and water

The themes of environment, land, and water have generated the richest body of knowledge relevant to mine closure, and little can be added to what have been well-documented subjects. The destruction of landscapes, despoliation of agricultural and forest lands, sedimentation and erosion, contamination, and pollution can have their roots to large volumes of mine wastes (van Zyl, Sassoon, Digby, Fleury, & Kyeyune, 2002) and are often enduring although they vary with the type of ore, mining method, system of ore processing, effectiveness of water management, among other factors, and – after mining operation stops – the overall nature of mine closure (Wireman, 2000). Considering that the decommissioning of mines has long been acknowledged as an important procedure involving engineering and technical activities, an immense body of knowledge is available that informs theory and policy on remedies to address environmental problems associated with post-mine operation. Existing literature deals with altered landscapes, loss of soil, ground contamination, and remedies to address mine-associated degradation of ecosystems. Several studies and initiatives have paid attention to the treatment and neutralisation of acid rock drainage, and the long-term effects of toxic chemicals and mineral sediments in water bodies. There has been significant interest in landform designs and reconstruction to address ground stability concerns.[3]

In many rural communities, 'water', 'land', and the 'environment' are not only bio-physical resources in their everyday life but incorporate a whole range of other dimensions: political, economic, cultural, and spiritual.

Planning for closure as part of life-of-mine

Using knowledge fed by literature of the first strand, a second group of literature informs theory and practice on mine closure as an intrinsic operational phase within the life of a mine. The due emphasis is on how to close a mine systematically as early as possible and adopt a beginning-to-end perspective during feasibility stage or commencement of operations. The aim is to develop strategies to ensure work sites are left in a condition that is compatible with desired land uses following mine completion. The literature along this line revolves around the 'Creed' of *planning well in advance* to minimise financial, social, and environmental risks in recognition of the underlying uncertainties that may emerge during a mine's operation until it eventually shuts down, and decades beyond cessation of operations. However, the fixation on *new* and *future* mines, which makes planning start off an empty or clean canvas, rather restricts what mine closure strategies can further accomplish. Also, the literature exists within a corporate genre where the ideal is to reduce all activity to a series of technical processes.

Support systems for mine closure

The third strand of scholarship has looked at systems for setting up workable frameworks for addressing mine closure concerns. These include legal/

legislative, regulatory, and financing aspects in which governments and financing institutions require mining operations to post reclamation bonds, also known as 'environmental bonds' or 'environmental guarantee fund'. The World Bank Group (Oil, Gas and Mining Policy Division, 2008) leads in the information on this subject, and special networks[4] also provide respected guidance. A few materials (for example, Fletcher, Hutton, & Dick, 2012; Gerard, 2000; Parshley, Bauman, & Blaxland, 2009) on this subject inform policy and mechanisms to ensure companies have the funds to deliver site reclamation following closure. Authors provide recommendations on cost estimations including a broad (e.g., national/federal) uniform approach (Woolley & Hutton, 2009). The book *Mining for Closure* by Peck et al. (2005), while intended for policy use within Southeastern Europe and within the Tisza River Basin, offers essential and broader principles in financing and implementing closure activities and rehabilitation for mines including abandoned and orphaned sites.

Social and economic aspects

The fourth strand of literature deals with the socio-economic effects of mine closure, an area which Warhurst and Noronha (2000, p. 14) have, earlier on, acknowledged as needing more research. Until the present, research along this direction is limited, particularly on how communities attempt to chart post-closure or post-mining directions. What is available is a variegated literature dealing with the social and economic repercussions of the closing down of mine operations on cities, towns, and remote regions. These include difficult adjustments to sudden collapse of economic activities, social change within communities, and economic and social dislocations of workers and communities (Bluestone & Harrison, 1982; Freudenburg, 1992; Machlis, Force, & Balice, 1990). Available accounts in Australasia demonstrate that the negative social impacts of mining persist long after mine closure with several of them completely irreversible. The Ok Tedi mining project (Jackson, 1993) in the Western Province of Papua New Guinea highlights this. The process of addressing the project's socio-environmental impacts exhibits the involvement of numerous social actors outside the mine vicinity, notably the Australian media and the Australian legal firm that represented the affected indigenous peoples in a legal battle against the mine (Banks & Ballard, 1997). In Britain, some studies focused on the economic dislocation resulting from the closure of coal mines in the 1980s. Migration had not been a significant response to the decline of the coal mining industry because mine workers are not being absorbed into the labour market in other ways (Hollywood, 2002). Conversely, a case in Africa, following the closure of Rio Tinto's nickel mine in 1982 in Zhombe, Zimbabwe, demonstrates that non-national workers were repatriated, but local workers were redeployed to regional and sub-regional branch networks. Other retrenched workers settled in Zhombe and integrated themselves with the community instead of going back to their places of origin (Madebwe, 2004).

A few studies dealt with activities to help ease or soften the adverse effects of mine closure, particularly loss of jobs. For example, government and non-government arrangements offered provisions for technical advice, guidance, and financial incentives to employees suffering the effects of an industrial plant relocation or closure at the Marmoraton Mining Company in Ontario, Canada in 1978 (Hegadoren & Day, 1981). However, much is still to be desired about the effectiveness of the initiatives.

A study focusing on post-mine transformation of regions where mining used to be the major – if not the sole – economic base, points to amenity-based growth as demonstrating the transformation of rural space and rural culture. Tourism and recreation as new forms of development have kept the link between community and natural resources (Winkler, Field, Lulloff, Krannich, & Williams, 2007). The study is one of few materials that offer insights into directions towards post-mining futures. Meanwhile, one study by Harner (2001) examines the process by which a physical space gains meaning as a 'place' through human experience and struggle. The people in two copper mining settlements in northeastern Sonora (Mexico) assume shared understanding about themselves which are interwoven with place meaning. While both settlements were under the same copper mining company, the local histories and social environments of each create diverging social realities, which in turn shape the process of identity construction. Thus, even with the influences of global forces in mining, local conditions and historical contexts interacting with landscape, configure the character and social space of a place. The place gains some symbolic significance that form part of the identity that people assume.

In a collection of case studies from Australia, Canada, Finland, Sweden, and Norway, Neil, Tykkyläinen, and Bradbury (1992) provide some understanding of the different factors that influence how affected people respond to threatened or actual mine closure. They recommend mining companies and governments strategies to address closure crises: developing schemes for assisting displaced workers in job search and retraining, preparing to cope with mine closures and downsizing, and employing expert consultants to assist the community in assessing the scope for local economic development. Expectedly, the volume presents settings which are different from many developing countries where governments often lack policy preparedness, much less financial resources (e.g., to employ expert consultants), to address socio-economic changes during the closure and post-closure periods.

The gap to fill

Environmental scientists, hydrologists, geomorphologists, geochemists, plant physiologists, environmental toxicologists, and many others work with planners, cost estimators, and other experts in highly specialised fields, as they apply intelligent thought and insightful innovations to mine closure plans and associated activities. They place emphasis on designing post-mining landscapes, assessing landform stability, analysing composition of mine waste materials, assessing risks and off-site impacts, predicting hydrological processes, and so forth – all towards restoring degraded landscapes and ecosystems. Undoubtedly, the step-by-step planning principles and guidance to ensure adequate financial provision for closure, environmental protection, usable land for the subsequent users, etc. towards creating a positive mining legacy are remarkably valuable and pragmatic principles that we must heed to. Nonetheless, the fixation on an 'empty canvas' which is emphasised in mine closure planning and is only possible for new and future mines – and entangled in the preoccupation with the minutiae of planning - create a disconnect from the immediate and everyday concerns of people and the reality that human beings are the be-all and end-all of any closure plan. Any purportedly well-planned mine closure undertaking that leaves out the historical, political, and socio-cultural elements embedded in communities' sense of being and belonging is not workable.

It is well-recognised that the integration of social and environmental elements in mine closure planning is crucial, and the participation and engagement of multiple stakeholders is imperative (Fleury & Parsons, 2006; Godlewski & Brown, 2010; Kamuzora, 2010; Pollet, 2009; Ross & Bond, 2008). However, existing recommendations and proposals pertinent to the 'social' component in mine closure planning have been limited to conducting 'consultation' and 'community engagement' with affected (or would-be affected) groups, and involving communities in planning and implementation of actions. On the ground, a mining company often deals with more than one type of community. As demonstrated in the cases that this book presents, communities are not homogenous. Even within a single type of community or a small group, members do not always speak with one voice and collective input or action is not necessarily available or attainable. In the literature on mine closure, some discussions make references to 'community ownership', 'community asset', and 'community viability' but scarcely bring to light the diverse sub-groupings, the intra- and interrelations including vehement disagreements between and among groups or clusters.

Compounding the inadequacy of understanding the intricacies within communities is the problematic notion of 'best practice'. Companies are eager to disseminate and replicate best practice in mine closure. However, it is notable that the mine operations that have been, so far, studied and documented as adopting and implementing best practices are mostly located in North American and European countries (Eger, Melchert, & Wagner, 2000; Warhurst & Mitchell, 2000). Central to ideas of 'best practice' is knowledge transfer from one 'in-the-know' to another (e.g., the community) that is deemed needing skills or knowledge. Mine closure practitioners must be cognizant that what may be regarded as a lack of knowledge among people in the community may in fact be a purposeful application of it. Working with other social actors must not be in recondite terms inherent in the idea of 'development' which imply the image of more knowledgeable outsiders giving help to the ostensibly powerless and less astute local people (Long, 1999). The suggestion of H. D. Smith (2008, 2009) of Australia's Northern Land Council, becomes relevant here. He notes that when it comes to formulating criteria for mine closure success, one must go beyond definitions and metrics of Western-scientific viewpoint. He strongly suggests paying attention to understanding the cultural and social reference points (Smith, 2008) and argues that planners must provide 'context, cultural value and significance to each specific location identified within any mining lease' which are available in local historic knowledge and mythology. By integrating a complementation of scientifically justified ecological measures and interpretive cultural indices, mine closure practitioners can develop useful, measurable, more meaningful cultural criteria. This requires going beyond fixation on 'how to' schemes, in the context of implementing 'best practice' or 'leading practice.' Luke Danielson (2005, p. 14), the chairperson of the two-year Mines, Minerals and Sustainable Development study, pithily summarises the knotty idea of 'best practice', and thus the need for grounding on locality-based circumstances:

> What may be a great technical solution to a tailings dam problem in the Atacama where it rains a few millimeters every ten years might be a disaster in Indonesia where it rains five meters of more a year. A state of the art approach to community consultation in Montana might, if repeated step by step, be a total fiasco in the Philippines.

An actor-oriented approach to the mine closure process allows the practitioners (e.g., mine closure personnel, community development workers, etc.) to delve into the lifeworlds of the other social actors. Enriched with ethnography, an actor-oriented approach includes probing the extent to which certain knowledge, including our own, are influenced by power relations and domains in which they are produced and located (Long, 2004). It allows going beyond the preoccupation with risk aversion, and exploring other actors' motivations and long-term aspirations, thereby asking the question 'best practice from whose viewpoint?'

Notes

1 Industry leaders have recognised the mineral industry's inauspicious reputation such that they formed in October 1998 the Global Mining Initiative, a program of the world's largest mining companies, to consider the social issues confronting the mining sector and, subsequently, determine what necessary reforms to implement. This initiative led to undertaking the Mining Minerals and Sustainable Development (MMSD) Project, a two-year series of studies launched in 2000 under the aegis of the World Business Council for Sustainable Development and carried out by the London-based International Institute for Environment and Development. The MMSD output, *Breaking New Ground*, while acknowledging the need for reforms within the industry, emphasises resolutely the clear need of societies for mineral commodities, and the inevitability for the industry to carry on (MMSD, 2002).

2 See Peck et al. (2005).

3 Numerous related studies along this line have been published in journals such as *International Journal of Rock Mechanics and Mining Science & Geomechanics*, and *Tunnelling and Underground Space Technology*.

4 Lawyer networks such as the Rocky Mountain Mineral Law Foundation (www.rmmlf.org) in Westminster, Colorado, USA is an example of such networks as it offers advice on financial guarantees of technical and environmental performance in the extractive industry.

References

AGDIIS (Australian Government Department of Industry Innovation and Science). (2016, September). Mine Closure. *Leading Practice Sustainable Development Program for the Mining Industry*. Retrieved from https://industry.gov.au/resource/Documents/LPSDP/LPSDP-MineClosureCompletionHandbook.pdf

Banks, G., & Ballard, C. (1997). Introduction. In G. Banks & C. Ballard (Eds.), *The Ok Tedi Settlement: Issues, Outcomes and Implications* (pp. 1–26). Canberra: National Centre for Development Studies and Australian National University.

Bluestone, B., & Harrison, B. (1982). *The Deindustrialization of America: Plant Closing, Community Abandonment, and the Dismantling of Basic Industry*. New York: Basic Books.

Danielson, L. J. (2005). Reviewers' Comments on the Framework for Responsible Mining. Retrieved from www.frameworkforresponsiblemining.org/pubs/LukeDanielsonComments.pdf

Eger, P., Melchert, G., & Wagner, J. (2000). Using passive treatment systems for mine closure – a good approach or a risky alternative? *Mining Engineering*, *52*(9), 78–83.

Fletcher, D., et al. (2012). The 'smoking gun' of detailed mine closure cost over-run – A review using case studies of the real costs associated with the demolition and removal of infrastructure in mine closure. In *Life-of-Mine Conference* (pp. 325–332). Melbourne: The Australasian Institute of Mining and Metallurgy.

Fleury, A. M. and A. S. Parsons (2006). Integrated Mine Closure Planning. In A. Fourie and M. Tibbet (Eds.), *Mine Closure 2006: Proceedings of the First International Seminar of Mine Closure* (pp. 221–225) Perth: Australian Centre for Geomechanics.

Freudenburg, W. R. (1992). Addictive economies: Extractive industries and vulnerable localities in a changing world economy. *Rural Sociology*, *57*(3), 305–332.

Gerard, D. (2000). The law and economics of reclamation bonds. *Resources Policy*, *26*(4), 189–197.

Godlewski, D. W. and M. J. Brown (2010). Community Planning and Interaction – Pend Oreille Mine Case Study. In A. Fourie, M. Tibbet and J. Wiertz (Eds.), *Mine Closure 2010: Proceedings of the Fifth International Conference on Mine Closure* (pp. 181–186.). Perth: Australian Centre for Geomechanics.

Harner, J. (2001). Place identity and copper mining in Sonora, Mexico. *Annals of the Association of American Geographers*, *91*(4), 660–680.

Hegadoren, D. B., & Day, J. C. (1981). Socioeconomic mine termination policies: A case study of mine closure in Ontario. *Resources Policy*, *7*(4), 265–272.

Hollywood, E. (2002). Mining, migration and immobility: Towards an understanding of the relationship between migration and occupation in the context of the UK mining industry. *International Journal of Population Geography*, *8*, 297–314.

Jackson, R. T. (1993). *Cracked Pot or Copper-Bottomed Investment? The Development of the Ok Tedi Project 1982–1991: A Personal View*. Townsville: Melanesian Studies Centre, James Cook University of North Queensland.

Kamuzora, M. F. (2010). Proactive Stakeholder Involvement – Tulawakas's Key Towards Successful Closure. In A. Fourie, M. Tibbet and J. Wiertz (Eds.), *Mine Closure 2010: Proceedings of the Fifth International Conference on Mine Closure* (pp. 187–192). Perth: Australian Centre for Geomechanics.

Laurence, D. C. (2006). Why Do Mines Close? In A. Fourie and M. Tibbet (Eds.), *Mine Closure 2006: Proceedings of the First International Seminar of Mine Closure* (pp. 83–94). Perth: Australian Centre for Geomechanics.

Long, N. (1999, October). The Multiple Optic of Interface Analysis. Retrieved from http://lanic.utexas.edu/project/etext/llilas/claspo/workingpapers/multipleoptic.pdf

Long, N. (2001). *Development Sociology: Actor Perspectives*. London and New York: Routledge.

Long, N. (2003). An actor-oriented approach to development intervention. In D. A. Cruz (Ed.), *Rural Life Improvement in Asia* (pp. 47–61). Tokyo: Asian Productivity Organization.

Long, N. (2004). Actors, Interfaces and Development Intervention: Meanings, Purposes and Powers. In T. Kontinen (Ed.), *Development Intervention: Actor and Activity Perspectives* (pp. 14–36). Helsinki: Center for Activity Theory and Developmental Work Research and Institute for Development Studies, University of Helsinki.

Long, N., & Long, A. (Eds.). (1992). *Battlefields of Knowledge: The Interlocking of Theory and Practice in Social Research and Development*. London and New York: Routledge.

Machlis, G. E., Force, J. E., & Balice, R. G. (1990). Timber, minerals, and social change: An exploratory test of two resource-dependent communities. *Rural Sociology*, *55*(3), 411–424.

Madebwe, C. (2004). The impact of mine closures on rural population dynamics: The case of Zhombe in Kwekwe District, Midlands Province, Zimbabwe. *Eastern Africa Social Science Research Review*, *20*(2), 59–70.

Mining Minerals and Sustainable Development (MMSD). (2002). *Breaking New Ground: The Report of the Mining, Minerals and Sustainable Development Project*. London: Earthscan Publications.

Mulvey, P., Baker, A., & Scott, P. (2012, September). Mine Closure and Waste Responsibilities and Liabilities. Retrieved from www.environmentalearthsciences.com.au/wp-content/uploads/2013/01/Mine-Closure-Paper.pdf

Neil, C., Tykkyläinen, M., & Bradbury, J. (Eds.). (1992). *Coping with Closure: An International Comparison of Mine Town Experiences*. London: Routledge.

Parshley, J. V., et al. (2009). An evolution of the methods for and purposes of mine closure cost estimating. In A. B. Fourie and M. Tibbett (Eds.), *Mine Closure 2009 Proceedings of the Fourth International Conference on Mine Closure* (pp. 187–200). Perth: Australian Centre for Geomechanics.

Peck, P., Balkau, F., Bogdanovic, J., Sevaldsen, P., Skaalvik, J. F., Simonett, O., Daussa, R. (2005). Mining for Closure: Policies and Guidelines for Sustainable Mining Practice and Closure of Mines. Retrieved from http://enrin.grida.no/environment-and-security/mining_src.pdf

Pollet, E. A. (2009). Mine Closure – Participatory Stakeholder Engagement and Socio-economic Closure Planning for Community Sustainability in the Context of an Uncertain Future. In A. Fourie and M. Tibbet (Eds.), *Mine Closure 2009: Proceedings of the Fourth International Seminar on Mine Closure*. (pp. 417–428). Perth: Australian Centre for Geomechanics.

Pring, G. R., Otto, J. M., & Naito, K. (1999). Trends in international environmental law affecting the minerals industry. *Journal of Energy and Natural Resources Law*, *17*(1), 39–55.

Rhomberg, C. (2004). *No There There: Race, Class, and Political Community in Oakland*. Berkeley, CA: University of California Press.

Ross, N. and B. Bond (2008). Viability of Aboriginal Communities Beyond Mine Closure. In A. Fourie, M. Tibbet, I. Weiersby and P. Dye. (Eds.), *Mine Closure 2008: Proceedings of the Third International Seminar on Mine Closure* (pp. 873–878). Perth: Australian Centre for Geomechanics.

Smith, H. D. (2008). Using Traditional Ecological Knowledge to Develop Closure Criteria in Tropical Australia. In A. B. Fourie, M. I. Tibbett, I. M. Weiersbye and P. Dye (Eds.), *Mine Closure 2008: Proceedings of the Third International Seminar on Mine Closure* (pp. 47–56). Perth: Australian Centre for Geomechanics.

Smith, H. D. (2009). Strangers in a Foreign Land - Developing Cultural Closure Criteria for Mines in Australia's Northern Territory. In A. Fourie and M. Tibbet (Eds.), *Mine Closure 2009: Proceedings of the Fourth International Conference on Mine Closure* (pp. 3–11). Perth: Australian Centre for Geomechanics.

van Ufford, P. Q., Kumar Giri, A., & Mosse, D. (2003). Interventions in development: Towards a new moral understanding of our experiences and agenda for the future. In P. Q. van Ufford & A. Kumar Giri (Eds.), *A Moral Critique of Development: In Search of Global Responsibilities* (pp. 1–40). London and New York: Routledge.

van Zyl, D., Sassoon, M., Digby, C., Fleury, A.-M., & Kyeyune, S. (2002). Mining for the Future Main Report. *Mining, Minerals and Sustainable Development*. Retrieved from www.iied.org/mmsd/mmsd_pdfs/draft_paper_mc.pdf

Warhurst, A., & Mitchell, P. (2000). Corporate Social Responsibility and the Case of Summitville Mine. *Resources Policy, 26*(2), 91–102.

Warhurst, A., & Noronha, L. (2000). Integrated environmental management through planning to closure from the outset: The challenges. In A. Warhurst & L. Noronha (Eds.), *Environmental Policy in Mining: Corporate Strategy and Planning for Closure* (pp. 13–32). Boca Raton, London, New York and Washington, DC: Lewis Publishers.

WB-IFC (World Bank and International Finance Corporation). (2002). It's Not Over when It's Over: Mine Closure Around the World. *Mining and Development Series.* Retrieved from www.natural-resources.org/minerals/CD/docs/twb/mine_closure.pdf

Winkler, R., Field, D. R., Lulloff, A. E., Krannich, R. S., & Williams, (2007). Social landscapes of the Inter-Mountain West: Comparison of 'Old West' and 'New West' communities. *Rural Sociology, 72*(3), 478–501.

Wireman, M. (2000). Hardrock mining. In G. E. Dissmeyer (Ed.), *Drinking Water from Forests and Grasslands: A Synthesis of the Scientific Literature* (pp. 179–186). Asheville, NC: United States Department of Agriculture Forest Service Southern Research Station.

Woolley, M., & Hutton, A. (2009). *New Regulatory Approaches in Rehabilitation Cost Estimation: How Successful Have They Been?* Paper presented at the Mine Closure 2009 Proceedings of the Fourth International Conference on Mine Closure, Perth, Australia.

World Bank Group (Oil, Gas and Mining Policy Divison), T. (2008). Guidance Notes for the Implementation of Financial Surety for Mine Closure. Retrieved from http://siteresources.worldbank.org/INTOGMC/Resources/financial_surety_mine.pdf

World Bank Multistakeholder Initiative. (2010, March). Towards Sustainable Decommissioning and Closure of Oil Fields and Mines: A Toolkit to Assist Government Agencies Version 3.0. Retrieved from http://siteresources.worldbank.org/EXTOGMC/Resources/336929-1258667423902/decommission_toolkit3_full.pdf

2 The socio-political landscape of mining in the Philippines

> We have rich resources under the earth that need to be extracted. Leaving them undeveloped render them valueless, but developing them needs capital. For this, we invite the investors. However, we must ensure that benefits from mining must accrue to the Filipino people particularly the indigenous communities. It is necessary and paramount that people see and experience for themselves the benefits from mining. For their part, the mining companies must demonstrate that lands devastated by mining can be rehabilitated. People are going against mining because of the industry's sins of the past. The Mining Act, unlike our previous laws, has the necessary provisions to absorb the environmental and social costs of mining. But we must give the Mining Act a chance before we listen to critics saying, 'Repeal the Mining Act' or 'No to mining.'[1]
>
> Engr. Eduardo Austria, Mining Operations Officer, Mines and Geosciences Bureau-Cordillera Administrative Region

This chapter outlines the historical and political-economic setting within which the minerals industry of the Philippines is embedded. The chapter provides a history of mining in the country focusing on Benguet Province where industrial mining began. A discussion on the saga of the Mining Act of 1995 is necessary to present the major social actors, diverse interests, and processes that highlight intertwining concerns and challenges which characterise the terrain of mining in general. Recent policy responses to public demands add to the overall backdrop of mining that in turn shapes industry mechanisms to address convoluted concerns associated with mine closure.

Location and mineral endowment

Comprising 7,107 islands, the Philippine archipelago lies within a geological zone of frequent earthquakes and intense volcanic activity at the borders of the tectonic plates. The area extends from the Philippine archipelago to

Indonesia, encircling the basin of the Pacific Ocean. Several authors (e.g., Domingo, 1993; Lopez, 1992; Symon, 2007; Walker, 1992) point to this location of the Philippines along the Pacific Ring of Fire as an explanation for the country's significant mineral wealth.

Data from the Mines and Geosciences Bureau (MGB) (2017c) indicate that out of the 300,000-square-kilometre total land area of the Philippines, 90,000 square kilometres – or 30 percent – has been identified as having high mineral potential. In June 2016, the area that is covered by mining tenements stood at 2.70 percent of total area or 8,100 square kilometres, although MGB notes that this expanse is still subject to the 'mandatory relinquishment by contractors'. The main commodities are copper, gold, and nickel, which are also the country's leading mineral exports (MGB, 2017b, 2017c).

Mining during pre-colonial era and Spanish regime

Before the colonialists arrived in the 1500s in what is now the Philippines, a flourishing trade existed between the islands' inhabitants and the Chinese (Ang See, 1990; Caballero, 1996; Lopez, 1992). This trade involved bartering gold ingots and granules from the islands with finished Chinese items such as jewellery, belts, and masks. According to Caballero (1996) the earliest site in the Philippines where gold artefacts are found is from as early as 400—250 B.C. In Lepanto[2] at the northern part of what is now the province of Benguet, a bronze-copper culture developed among the Kankanaey who exchanged copper and iron wares with livestock, salt, dried fish, mats, beads, jars, blankets, and other manufactured goods with the coastal towns of the Ilocos. Similarly, the Ibaloy in the southern part of Benguet traded their cattle, horses, and processed gold with lowland products. It was through this trade that the Spanish became aware of the gold mines (Bagamaspad & Hamada-Pawid, 1985).

During the late 1500s and 1600s, the Spaniards sent frequent expeditions into northern Luzon to take control of the gold mines which the Kankanaey and Ibaloy worked (Phelan, 1959; Scott, 1974; Zuniga, 1966). According to the Cordillera scholar W.H. Scott, 'it was the insult to His Majesty's purse rather than to his pride which led to the so-called conquest – or at least devastation – of Igorot Territory' (1970, pp. 714–715). Early Spanish accounts of the expeditions describe the widespread gold mining throughout Luzon and the role the metal played in the local economy (Caballero, 1996). Earlier referred to as *Ygolotes* by the Spanish chronicler Dr. Antonio De Morga[3] (1609, Re-published in 1907), the Igorots had rich mines and used gold and silver to trade with swine and buffalos with the Ilocos. They successfully repulsed the intruding strangers. Missionaries who visited the highlands without military escort often died at the hands of the fierce Igorots (Phelan, 1959). By the nineteenth century, following the series of unsuccessful military pursuits which incurred huge costs to the Royal Treasury, the Spanish finally ceased attempting to seize the Igorot gold mines (Scott, 1970, 1974).

Towards the mid-1800s, the Spanish colonial government legislated the institutional regulation of the mining industry. By virtue of conquest, the entire archipelago of the Philippines was assumed to be owned by the Spanish King (Prill-Brett, 1994). This assumption, referred to as the Regalian Doctrine, although considered by some legal analysts as 'mythical'[4] (Leonen, 2004, p. 153), hinges on the principle of eminent domain which accords the Crown the right to develop mines on its own initiative or through private concessions. The Regalian Doctrine however was not extensively effective during the Spanish regime because the colonial government was not able to subjugate all areas of the islands. Nonetheless, the doctrine would later become the basis upon which subsequent Philippine land laws proceed (Lynch, 1986).

Beginning of industrial mining

In what is now present-day Benguet Province, it was under the American rule when newcomers took hold of the mines and began exploitation by modern technology. These include former servicemen engaged in gold mining in Alaska and California but remained in the Philippines after the Spanish-American War in 1898. The adventures of these soldiers-turned-prospectors involved searching for gold not through finding geologic nodes. Rather, they found gold by establishing friendship and living with the Igorot people who adorned themselves with gold necklaces, earrings, rings, and other ornaments (Lopez, 1992). Some of them, particularly in Mankayan in the northern part of Benguet, married Kankanaey women (Bagamaspad & Hamada-Pawid, 1985).

The American colonial government introduced land registration in which communal resources such as minesites became private properties. The United States Congress enacted the Public Land Act of 1902, also known as the Philippine Bill of 1902, which declared minerals in public lands as free and open for exploration, occupation, and purchase by citizens of both the United States and the Philippines. This law led to the application for patents and land titles over mineral lands. However, among Filipinos, only a few, mainly the elites and schooled individuals, availed of the land registration system (Keesing & Keesing, 1934; Prill-Brett, 2003; Prill-Brett & Salinas-Ramos, 1998). The law unwittingly disenfranchised the great majority of the inhabitants of the Philippines, most of whom found the idea of a documented land title alien and incomprehensible (Lynch & Talbott, 1995).

During my fieldwork in mid-2003 and 2004–2005, narratives and recollections about *deklarasyon* (land registration) often came up during my conversations with research participants in Itogon. A local council official of Barangay Ampucao narrated:

> Large areas of land in Itogon have been controlled by mining companies for a long time. According to my parents, their old folks unmindfully permitted white men and fellow Filipinos to come around to survey lands. Little did our old folks realise that the surveying of lands was a move towards registration. The old folks just found out one day that certain individuals – of strange names, because they were not from Itogon – already own the mines. In some instances, some people may have given the permission for those men to only do some work underneath the ground where houses stood. Eventually, the naïve old folks found out that the ground both underneath and surface is already owned by someone else.[5]

The introduction of land documentation accentuated the incongruity between colonial concepts of property and unwritten indigenous land tenure systems. In Mankayan in northern Benguet, the marriages of some American mining prospectors to local Kankanaey women led to the maintenance of a few economically prominent families who were able to own mining claims in the *Sociedad Minero Metalurgica Cantabro-Filipino*, which was restored in 1907, to become the present-day Lepanto Consolidated Mining Company (LCMCo.) (Bagamaspad & Hamada-Pawid, 1985). LCMCo. became the new owner of the lode and placer sites traditionally worked by many Kankanaey. Other families left Mankayan due to their new status as 'squatters' on surface areas which the new owners must now protect. Departing from Mankayan in the north and moving to gold-rich Itogon in southern Benguet, they looked for places where to reside and lands where to mine. Many of them settled in Dalicno, Tuding, Ucab, and areas within the present-day barangays of Ampucao and Gumatdang. Meanwhile, the Ibaloy in Itogon also lost their lands to various individuals becoming the new owners of minesites. The American company Benguet Consolidated took control, by registration, of extensive areas of land in Itogon. Many indigenous Ibaloy left their original minesites in Gumatdang, Acupan, and Antamok to move into the lower Agno River (Bagamaspad & Hamada-Pawid, 1985).

The exhilaration of prospecting and grubstaking in Itogon was accompanied by ongoing developments in Baguio, a pine-clad plateau with temperate climate, into a hill station where exhausted and injured American soldiers could recover. The completion of the winding 40-kilometre Kennon Road in 1905 made Baguio accessible to colonial bureaucrats and prospective mining grubstakers (Lopez, 1992). Some of these men became founding members of the elite Baguio Country Club that built a clubhouse offering facilities for various sports such as croquet, trap shooting, tennis, and golf, thus becoming the social centre of the city (Baguio Country Club, 2015). Migration into Baguio was immediate and continuous with the additional allure of nearby Itogon to the American capitalist (Lopez, 1992).

Mining policies before the Mining Act of 1995

Following the land titling laws enforced by the Americans that saw mining claims being staked openly, more incentives for mining rebates were enacted by Congress to help offset production costs. Under President Ferdinand Marcos, a smooth path for the extraction of the country's mineral resources had been laid out with specific laws created for such purpose (Malig, 2002). In 1971, the Senate and Congress enacted Republic Act 6364, also known as the Gold Subsidy Law, to provide relief to gold producers by requiring the Central Bank to purchase newly mined gold from mining companies at a designated price (RA 6364, 1971).

In 1974, Marcos issued Presidential Decree (PD) 463 to provide for an efficient administration and disposition of mineral lands and promote and encourage their development and exploitation. Under the law, mining companies were exempt from paying customs duties and all taxes for machineries, equipment, tools for production, and plants imported for the use of new mines and old mines. In addition, all mining claims, improvements, and mineral products derived from these claims were not liable for the payment of all taxes except income tax (PD 463, 1974, Chapter X, Section 53). PD 463 is the country's first mining law to provide for mechanisms to address pollution and waste problems associated with mining (Chapter XIV, Section 81), as well as social and economic development for communities particularly during the post-mining phase (Chapter XII, Section 91). However, the law concurrently bestowed mining companies timber, water, and easement rights on mining claims they own, occupy or lease (Chapter XI, Sections 56–58). While the law stipulated penalties for pollution from mine wastes and mill tailings, the maximum fine of PhP5,000 or six-year imprisonment or both (Chapter XIV, Section 81) was never enforced on polluting mines. PD 463 also aimed to open the opportunity for local investors to engage in mining. However, this opportunity was given to only a few elites (Malig, 2002). Whose ownership of the majority shares were complemented by several incentives like a lease term of 25 years, renewable for another 25 (PD 463, 1974, Chapter VIII, Section 40). During the mid-1970s, additional laws were issued by Marcos that all aimed to protect the environment. Presidential Decree 984, also known as the Pollution Control Law, issued in 1976, was one of several impressive laws.

Following the expansion of Marcos's powers after he declared martial law, there was a drastic increase in international borrowing for huge development projects (Howard, 1988, p. 72). When the Philippine economy slowed down in the late 1980s, many major mining firms collapsed. The collapse was traced to the excessive protection that the government provided for businesses which relied broadly on 'corruption, fraud and pulling strings with those in power' (The Economist, 1983, p. 71).

Marcos's successor, Corazon Aquino, issued Executive Order (EO) No. 279 (1987), as an interim law between the leasehold system until a new mining law and enabling regulations could be passed. EO 279 authorised the Secretary of the DENR to 'accept, consider and evaluate proposals from foreign-owned corporations or foreign investors for contracts or agreements involving either technical or financial assistance for large-scale exploration, development and utilization of minerals'. This role, which was previously a prerogative of only the president of the Philippines, effectively expedited the entry of many foreign firms into the country. Like Marcos, Aquino encouraged foreign investors to enter the industry by giving tax holidays, tax reduction of the first five years, and other tax exemptions (Malig, 2002). The constitutionality of EO 279 was challenged in 1990 but upheld by the Philippine Supreme Court in 1995 (Supreme Court of the Philippines, 1995).

Aquino's successor, Fidel V. Ramos, further opened the doors to foreign investors for all industries, deregulating, liberalising and privatising almost all government-owned corporations. After Corazon Aquino began economic restructuring which included liberalising the economy that involved privatising more than 250 government-owned and -controlled corporations (Gonzalez, 2001), Ramos built on these foundations by launching market-based economic reforms. As the political-economic analyst Walden Bello (2004, p. 92) would describe it, Ramos's scheme 'for liberalisation, deregulation and privatisation was pursued with almost messianic zeal'. In 1995, Ramos signed into law Republic Act 7942, also known as the Mining Act of 1995.

The Mining Act of 1995

The law that currently governs mining in the Philippines is Republic Act No. 7942 (1995). Its passage was the result of the lobbying by the Chamber of Mines of the Philippines (COMP)[6] for many years. It is a response to the directive of President Fidel V. Ramos to simplify 'the old executive controls' under which mining projects were being administered (The Mining Journal, 1995). Its provisions reflect the government's neo-liberal economic policies aimed at drawing investments. The law engendered controversies, particularly among nationalist groups, due to its provisions allowing up to 100 percent ownership of mining operations by foreign companies. Before the enactment of this law, foreign-equity ownership was pegged at a maximum of 40 percent. President F. V. Ramos described the Act as a 'powerful weapon for mass upliftment' in alleviating poverty and gives the country a competitive advantage to join the more advanced economies of Asia (Ramos, 2005). Ramos's successor Gloria Macapagal Arroyo[7] provided more impetus for the law through her policy of 'promoting mining while adhering to the principles of sustainable development' (Executive Order No. 270, 2004).

Barely a year after the Mining Act's enactment, 20 of the world's largest mining companies filed applications for various mining tenements and exploration projects in the Philippines (Cabalda, Banaag, Tidalgo, & Garces, 2002; Domingo, 2003). However, developments for many mining projects have been sluggish due to a composite of political, policy, technical, financial, and environment-related factors. The Mining Act has had almost a decade-long thorny trail (Chaloping-March, 2011) in which its supposed promise for wealth creation has hardly been proven. Several factors caused the law to take a back seat. Foremost of these is the unrelenting opposition to mining that is grounded in nationalist principles such as state sovereignty and national control over resources. In 1997, a group of NGOs, led by the the Legal Rights Center-Kasama sa Kalikasan (LRC-KsK),[8] challenged the constitutionality of the Mining Act. The petitioners of the case, known as the *La Bugal-B'laan Tribal Association, Inc. v. Ramos* (Supreme Court of the Philippines, 2004), argued that allowing 100 percent foreign ownership in large-scale exploration, development, utilisation, and exploitation of mineral resources in the Philippines through Financial or Technical Assistance Agreement (FTAA) breaches the constitutional provision that minerals in the Philippines are a national heritage of the Filipino people.

During the legal impasse, another law was enacted that has critical implications for approval processes of mining projects particularly within ancestral lands – the Indigenous Peoples Rights Act (IPRA) (Republic Act 8371, 1997). One of the COMP officers described the IPRA as 'the last nail on the coffin for the mining industry'.[9] Exploration companies were hesitant to commit any further capital to exploration activities due to the uncertainties brought by the IPRA (Disini, 1999). The IPRA's final passage on 29 October 1997 resulted from decade-long lobbying, deliberations, and consultations by concerned NGOs, Peoples Organisations (POs, and indigenous peoples' representatives with the support of public interest lawyers who are themselves environmental activists. IPRA provides for priority rights to indigenous cultural communities/indigenous peoples (ICCs/IPs) in the extraction, development or exploitation of any natural resources within their ancestral domain.[10] Soon after IPRA's enactment, two lawyers challenged at the Supreme Court the constitutionality of the IPRA. Their suit assailed the rights of ICCs/IPs over the natural resources within their ancestral domain, with one argument, among others, that it is an 'an unlawful deprivation of the State's ownership over lands of the public domain', minerals and other natural resources. On 6 December 2000, the Supreme Court dismissed the constitutionality challenge (Supreme Court of the Philippines, 2000). Subsequently, the petitioners filed a motion for reconsideration. However on 21 September 2001, the Supreme Court resolved the IPRA as constitutional.

Meanwhile, beginning the time La Bugal-B'laan case was filed in 1997, there had been no decision by the Supreme Court until it finally issued on January 2004 its ruling nullifying the Mining Act's FTAA provisions (Supreme Court of the Philippines, 2004). The MGB, the private respondents, and COMP immediately appealed the decision. Among other arguments, the president of the COMP noted that the Philippines has already lost at least US$20 billion worth of export revenues because the Mining Act was not fully implemented since its passage in 1995 (Clancy, 2005). Working closely with the Office of the Solicitor General, the COMP, on behalf of its members, filed a motion for reconsideration. An exhaustive constitutional review and an oral hearing were held in July 2004. On 1 December 2004, the Supreme Court reversed its January decision, affirming the legality of the FTAA and the Mining Act in its entirety. The LRC-KsK submitted a motion for reconsideration of the Court's ruling. However, on 1 February 2005, the Supreme Court denied the motion, thus upholding with finality the constitutionality of the FTAA provision and the Mining Act.

Mine tailings spills

In March 1996, barely a year after the issuance of the Mining Act just as the government was starting to revive the industry, a major tailings spillage from the Marcopper mine occurred (Plumlee, Morton, Boyle, Medlin, & Centeno, 2000). The immediate effects were catastrophic: agricultural fields were flooded and fishing which was a major livelihood for more than 20,000 families in 42 communities stopped due to the flow of mine tailings burying the channels and the valley floor (Pasimio, 2005). While the spillage was committed specifically by Marcopper, widespread denunciation by environment activists was launched at the entire mining industry. The disaster stirred doubts at the industry's claims of environmental management especially in handling mine wastes and tailings (Ramos, Cabalda, & Banaag, 2000).

In 2005, the new Rapu-rapu Polymetallic Project (RRPP) finally advanced to its commercial operation with its first production in July 2005. This was a significant milestone for the minerals industry because the RRPP was the first to have opened and eventually advanced into production in the country within three decades (MGB, 2005). Operated by Lafayette Mining

Philippines Incorporated (with the Sydney-based Lafayette Mining Limited as majority – i.e., 74 percent – shareholder), the RRPP was considered by the Philippine government as critical to attracting more global investments into mining in the Philippines (DENR, 2006; RFFC, 2006). Further, the project has been promoted as bringing development to the small and fragile Rapu-Rapu island through employment opportunities and community development projects such as building of roads, school houses, electricity, water systems, etc. (RFFC, 2006). Barely three months after RRPP's first gold production, mine tailings spilled from the company's ore processing mill into nearby creeks leading into the sea. In its assessment report on the incident, the DENR stated (2006, p. 7): 'The Lafayette project does not appear to measure up to the standards of responsible mining' and 'the project does not comply with the "best available practices" as advocated by Environment Australia, or even by the Code of Conduct of the Philippine Chamber of Mines.' In the same report, the DENR admitted its culpability 'in not being able to monitor' in advance before any incidents happened. It acknowledged its being 'dysfunctional' as it failed to check on the company's 'crucial non-compliance' to rules and regulations (DENR, 2006, p. 3). While Lafayette received regulatory sanctions, the tailings spillage caused a major blow to the Philippine government's campaign for mining as an economic base for communities and as an investment venture in the country. It added more weight to antagonism towards mining and fueled the already stiffening opposition by various civil society elements, including religious sectors. Lamentably, it also raised questions about the capability of the DENR as a regulating agency. In August 2012, another tailings spillage took place – this time, from the tailings storage facility of Philex Mining Corporation.

Recent policies since 2012

To enhance environmental mining standards and increase revenues from the minerals sector, President Benigno Aquino III issued Executive Order No. 79 (2012). The law added areas closed to mining applications and placed a moratorium on new mining agreements pending a new legislation rationalising existing revenue sharing schemes and mechanisms.

As part of raising environmental standards within the large-scale minerals sector, the DENR issued in 2015 a policy (DENR Administrative Order [DAO] No. 2015-07, 2015) requiring all mining contractors to secure ISO 14001 certification, particularly on environmental management systems. Members of the COMP complied and the majority have secured certification. Similarly, DENR also issued a separate policy (DAO No. 2015-03, 2015) ordering small-scale miners to organise themselves into cooperatives, register with their respective provincial/city mining regulatory boards, and thus apply for a *Minahang Bayan*, a concession area where they are permitted to mine. The board reviews applications for subsequent approval of the DENR secretary. The policy defines the responsibilities of small-scale miners including, among others, technical capability and the financial resources to meet the structural and taxation requirements of the concession area. At present, there are merely five[11] declared Minahang Bayan in the country.

Meanwhile, The Philippines became an EITI candidate country in May 2013 and the Philippines-EITI Multi-Stakeholder Group has already produced its Third Report (Ph-EITI, 2016). All mining contractors are now mandated to comply with the disclosure requirements of the Philippine Extractive Industries Transparency Initiative (Ph-EITI). Under President Rodrigo Duterte's Administration, the DENR Secretary Gina Lopez issued in 2017 a policy (DAO No. 2017-07) requiring all mining contractors to take part in Ph-EITI in accordance with the requirements set by Philippines Multi-stakeholders Group. This development is immensely significant as the public now has clearer information on the real contributions of the minerals industry to the country's economy, among others. The open reports and associated information, through Ph-EITI Multisectoral Group, provide opportunities for government, industry, and civil society to pinpoint concerns about resource management and governance systems, hopefully, towards addressing problems.

Major social actors

Large-scale corporate mines and small-scale operators

The minerals industry of the Philippines has two sectors. First is the large-scale corporate sector which includes companies that undertake huge mechanised industrial operations. Their volumes of production are sizeable enough to be commercially viable. Latest records of the MGB (2017a) indicate that the Philippines at present has 41 metallic mines which include three copper mines, six gold mines, 27 nickel mines, four chromite mines, and one iron mine.

The minerals industry also has the small-scale sector. Due to space considerations, this discussion concerns itself only with the 'small-scale' gold sector and does not cover other commodities. The sector includes scores of thousands of individuals and hundreds of loose miners' organisations as participants – in varying forms, capacities, and arrangements.[12] This sector employs about 300,000 miners who operate in 30 of the country's 81 provinces, and account for up to 70–75 percent of the gold production in the country (Mendoza, 2014), albeit the majority of participants do not have permits to operate. While large-scale mining companies were closing in the 1990s, the small-scale gold mining sector had been peaking up partly in response to the large-scale sector's decline. Small-scale gold mining became a socio-economic phenomenon beginning the 1980s with the discovery of gold rush site in Davao followed by scattered gold panning activities in 37 other provinces. Production from small-scale mining and panning activities during this time reached an unprecedented record of 11 tonnes, exceeding that of primary and secondary producers (MGB, 2004, p. 1)

The Department of Environment and Natural Resources (DENR)

The primary government agency that takes charge of the 'conservation, management, development, and proper use of the country's environment and natural resources' is the DENR. It is responsible for the 'assessment, development, protection, conservation, licensing and regulation as provided for by law, where applicable, of all natural resources' (Executive Order No. 192, 1987, Section 4 & Section 5, Paragraph k). Within the DENR is the Mines and Geosciences Bureau (MGB) which is directly in charge with the 'administration and disposition of the country's mineral lands and mineral resources' (DAO No. 1996-40, 1996, Chapter II, Section 7). Based on a report by the Asia Pacific Centre for Environmental Law (Tan, 1998), the Philippines is the only one among the Southeast Asian countries that places the mandates of environmental protection and natural resource management in a single agency, i.e., the DENR. Some observers and critics point out that while the agency is tasked to protect the environment, it is ironic that it also promotes mining. The MGB's response to this criticism is:

> We have been accused of being biased toward the mining industry. We are not. Our bias is toward RESPONSIBLE mining. It is our MANDATE – as the government agency responsible for the management and administration of the country's mineral resources, to promote and enhance responsible minerals resources development. We will defend the mining industry as long as it toes the line of Responsible Mining. However, we will not hesitate to crack the whip where principles and policies of Responsible Mining are breached.
>
> MGB (2000, p. 19, capitalisation in original)

Civil society groups

A major feature of the socio-political landscape of mining in the Philippines is the dynamism of civil society organisations. The NGOs and POs in the Philippines are extremely active and considered the largest in Asia in terms of per capita population (Wurfel, 2004). Due to their substantial role in development affairs within the Philippine society, many scholars have paid attention to their general makeup, activities, areas of operation, and growth (see Alegre, 1996; Constantino-David, 1992; Hilhorst, 2003; Miralao & Bautista, 1993; Serrano, 2002; Silliman & Garner Noble, 1998).

The development and vigour of civil society in the Philippines find support in the 1987 Philippine Constitution which declares that 'the State shall encourage non-governmental, community-based, or sectoral organisations that promote the welfare of the nation' (Article 2, Section 23). The Constitution also recognises that independent people's organisations are mechanisms by which people can 'pursue and protect, within the democratic framework, their legitimate and collective interests and aspirations through peaceful and lawful means' (Article 13, Section 15). Further, 'the right of the people and their organisations to effective and reasonable participation at all levels of social, political, and economic decision-making shall not be abridged' (Article 13, Section 16).

NGOs whose advocacies cover the interlinking themes of land rights, human rights, environmental protection, and anti-neoliberal economic development emerged and multiplied during the 1990s, in response to three events (Cabalda et al., 2002, p. 59) that had stacked up on each other towards a saturation point: a) the opening of an open pit mine in Antamok, Itogon during the 1990s, b) the passage of the Mining Act in 1995 that gives foreign-owned mining companies full entry to the Philippines, and c) the Marcopper tailings spillage in 1996. In particular, the third event served as a major catalyst for mining opposition (Ramos et al., 2000).

Policy making has been an avenue through which civil society groups have taken their rejection of the current mining policy. The 1987 Constitution created the party-list system to provide representation of the apparently under-represented sectors of the Philippine society to 'contribute to the formulation and enactment of appropriate legislation that will benefit the nation as a whole' (Republic Act No. 7941, 1995, Section 2). These sectors include labour, peasant, urban poor, indigenous cultural, women, youth, and other such sectors, except the religious sector, as may be defined by law. In August 2016, *Akbayan* Senator Ana Theresa Hontiveros Baraquel filed Senate Bill 1069 (2016), also known as the 'Alternative Minerals Management Bill'. The bill, which enjoys the strong support of community leaders and advocates from *Alyansa Tigil Mina* (2016),[13] aims to 'rectify the failure of Republic Act 7942'. The 'People's Mining Act of 2016', or House Bill 4315 (2016), was filed on 9 August 2016 by *Bayan Muna* Partylist Representative Carlos Zarate and Act Teachers Party Representative France Castro.[14]

Local government units

The Philippine State is subdivided into territorial-political units – namely the provinces, cities, municipalities, and *barangays* (villages) all of which form the local government units (LGUs). As of 30 June 2015, the Philippines has 81 provinces, 144 cities, 1,490 municipalities, and 42,029 barangays (NSCB, 2015). The province, being the largest unit in the political structure of the country, consists in varying numbers of municipalities, and in some cases, component cities. The Local Government Code (RA 7160, 1991) delegated certain functions, such as environmental management, to LGUs which include provincial, municipal and barangay levels of governance. The national government is the permitting and regulatory authority of the extractive industry but the LGUs play an important role in large-scale and small-scale mining operation within their territorial jurisdiction.

The convivial stance of the national government towards mining projects and the enthusiasm of the COMP in promoting the Philippines as a haven for mining investments have not been shared by several local governments (Chaloping-March, 2014). There have been varying degrees of local government inhospitality towards potential mining projects within their specific political-geographic jurisdictions. One LGU that demonstrated early unreceptiveness towards mining is the provincial government of Capiz that imposed in August 1999 a 15-year moratorium (up to 50 years as maximum) on large-scale mining, but was lifted in May 2016 (House Resolution No. 600, 2016). The staunchest opposition to mining by local governments, supported by their constituent citizens, have been waged in Palawan (MISN, 2010), Romblon, and South Cotabato (Minda News, 2011a, 2011b). Local ordinances are being challenged by mining companies and the MGB-DENR, both arguing that national government laws have primacy over local government laws. Together with local organisations, several LGUs thwart the entry of large-scale mining into their own jurisdiction on the grounds of environmental protection, nationalist sentiments, and equity issues.

The clergy

In a country of predominantly Catholic devotees, the role and influence of the Roman Catholic Church in social and political life cannot be overstated. Through the Catholic Bishops' Conference of the Philippines (CBCP), the stance of the Catholic Church on mining was expressed as early as 1998 through the official statement that calls for the repeal of the Mining Act (CBCP, 2006, 1998). In 2006, the CBCP called on the Arroyo government to cancel 'all concessions and deny all applications' (The Manila Times, 2006), and sought audience (which they were granted) with the president on their plea. In addition to the collective declaration by the CBCP, individual bishops have been crusading against opening mining, criticizing mining companies, and blaming the industry for deaths in natural disasters (Catholic News, 2006, 2007).[15] In 2006, Bishop Arturo Bastes led an initiative to investigate the Rapu-rapu tailings spill. His report was a substantial input into DENR's own assessment of the event. The bishop has reiterated his stance that the adverse effects of mining far exceed its benefits (Bicol Today, 2012). Similarly, the CBCP National Secretariat for Social Action, Justice and Peace led a civil society fact-finding mission (CBCP-NASSA et al., 2012) on the 2012 tailings spill of Philex Mining Corporation, and has called for the permanent closure of the company. Most recently, the CBCP joined hands with Alyansa Tigil Mina in urging Pope Francis 'to support and stand with the communities, local organizations and popular movements who are responding to the challenges of the mining industry'. The formal statement notes, 'We urge Pope Francis to support the call of Philippine Bishops and the mining-affected communities to repeal the Philippine Mining Act (RA 7942),' among others (Human Rights Online Philippines, 2015).

Communities in mining areas

Across the Philippines, villages and towns are affected by mining operations, some negatively and others positively, in varying degrees. Undoubtedly, mining projects had brought employment to thousands of people especially in former rural and hinterland areas that were transformed into booming towns. Examples of huge settlements established by mining companies in key mining provinces and municipalities include: Toledo City[16] (Cebu Province), Acoje[17] (San ta Cruz, Zambales), Lepanto (Mankayan, Benguet), Philex (Tuba, Benguet), Sangilo[18] (Itogon, Benguet), and the communities of Acupan, Antamok, and Balatoc (Itogon, Benguet). Both former and current residents of these communities have benefited from economic opportunities such as, among others, long-term employment, as well as elementary and secondary education for schoolchildren. Thus, it is not surprising that ardent support and testimonies about the positive socio-economic contributions of the industry to families, would come from these communities.

Equally, however, other communities bear the direct and indirect effects of tailings spills and other degradation associated with mining operations. Prior to the tragic 1996 Marcopper tailings disaster, the MGB recorded at least seven major environmental incidents involving mines covering the period 1992–1995 whose repercussions include inundation of farmlands, siltation, and coastal pollution (Ramos et al., 2000, pp. 13–15). They brought havoc to agriculture and fishing, two critical economic activities of rural villages and towns in the Philippines.

In the 1950s–1970s, the pinnacle of Philippine mining, there had been virtually no criticism and opposition against mining operations because the catastrophes associated with mining were not yet evident, hence they were less known. In the 1970s, several mines with marginal ore grades were opened, with companies having inadequate capital but recklessly investing in major projects, supported by guarantees from various government financial institutions. When mine closures took place, the government not only inherited huge debts but several mines with no rehabilitation, whose environmental risks until the present (Domingo, 2003) have served as bases for sustained opposition to mining.

Within Benguet province for example, former minesites had been left in the 1980s without any rehabilitation. These include Lobo[19] (Kibungan municipality but also extends to Atok and Kapangan), Santo Nino[20] (Tublay), and Camp Six[21] (Tuba). The duration of mining operation in these places was relatively short-term, and may explain why these villages and towns did not establish into larger economic hubs. Understandably, the people in these areas cannot claim and talk about social and economic prosperity that are spoken of by men and women who have grown up and lived in minesite communities that evolved with mining operations.

Notes

1 Author's interview, 18 March 2005, Baguio City. Research participant consented to have his name known.

2 Lepanto is mentioned in several Spanish accounts. Notwithstanding the changes in political boundaries, Lepanto as the traditional mining locality during the Spanish era has remained the same copper- and gold-producing district until the present.

3 Dr. Antonio de Morga held the position of auditor of the *Real Audiencia* (Court of Justice) of Manila. In his memoir *Sucesos de las Filipinas* (*The Philippine Islands*) he kept a chronicle of his observations on the Philippines and neighbouring countries during the sixteenth century.

4 Marvic Leonen, an expert in Philippine indigenous law, considers the Regalian Doctrine as 'a mythical and historically fallacious principle that permeates the thinking of, but has prompted little reflection within, the Filipino legal profession (Leonen, 2004, p.153).' He contends that the Regalian Doctrine is not supported by the legal and political history of the Philippines.

5 Author's interview, 24 April 2005, Ampucao, Itogon.

6 The Chamber of Mines of the Philippines is the aggregate of companies engaged in exploration, mining, and mineral processing. At present, it has 26 regular members and 14 associate members and also includes private law firms and professional industry associations.

7 In 1994, then Senator Gloria Macapagal Arroyo lobbied for a mining bill which would become the Republic Act 7942.

8 The Legal Rights Center-Kasama sa Kalikasan is an affiliate of Friends of the Earth International. It was a co-petitioner and co-counsel for members of the La Bugal B'laan people.

9 Author's interview with Nelia Halcon, November 2004, Pasig, Metro Manila.

10 In principle, no mining project whatsoever within ancestral domains of ICCs/IPs shall be allowed or renewed without going through the process provided by

law (NCIP Administrative Order 2012-03, 2012). However, the exercise of free prior informed consent (FPIC) by ICCs/IPs has remained extremely problematic. Huge challenges in terms of substance and procedure, among other factors, have continued to characterise the process of obtaining FPIC (by consent seekers) and expressing it (by consent givers).

11 The five *Minahang Bayan* are located in the provinces of Quezon, Eastern Samar, Davao Oriental, Agusan del Sur, and Dinagat Island.

12 Small-scale gold mining is not homogenous. It has several forms configured by varying historical-cultural and geographic-ecological circumstances. In the Cordillera region for example, entry and participation in the venture involve some criteria including locality and membership in the specific ethnolinguistic group in the mine area (Chaloping-March, 2015a, 2015b)

13 The *Alyansa Tigil Mina* (a Tagalog phrase denoting 'Alliance to Stop Mining') is 'both an advocacy group and a peoples' movement'. It includes NGOs, POs, church groups and academic institutions whose goal is 'to collectively challenge the aggressive promotion of large-scale mining in the Philippines' (2017).

14 The bill was earlier filed in 2011 by Representative Teodoro Casino in the 15th Congress and then refiled in 2013 by Representative Neri Colmenares in the 16th Congress as House Bill No. 171, which aimed to incorporate the 'Alternative Minerals Management Bill' authored by Ifugao Representative Teodoro Baguilat. The bill proposes substantial revisions including a change from the current export orientation of mining to a domestic needs-based development of the industry (CEC, 2011)

15 A strong network of NGOs in the advocacy of opposing mining is the DIOPIM Committee on Mining Issues (DCMI), based in southern Philippines. It is spearheaded by the Roman Catholic Church and composed of directors of Catholic social action centres, heads of other NGOs and indigenous people's groups. DIOPIM aims to 'to liberate people from the belief that mining promotes sustainable development' (DCMI, 2006).

16 Toledo City city developed with the Atlas Consolidated Mining which closed in late 1980s. In 1994, the company suspended its operations indefinitely but resumed in 2008 (http://www.atlasmining.com.ph)

17 Acoje developed through the operation of Acoje Mining Company, organised in 1935, and had produced chromite for more than 75 years. By 2010, other companies took over the tenement to mine nickel.

18 Sangilo is the community that developed with the former Itogon-Suyoc Mining Inc. which shut down operation in the early 1990s.

19 Lobo was the location of Western Minolco Mining operation which was incorporated in 1970, suspended in 1977, then resumed in 1979 but closed in 1982.

20 Santo Nino was location of the open pit mine operation of Santo Nino Mines which commenced operations in 1972 but closed in 1982 due to uneconomic ore, low metal prices, and insufficient capital.

21 Camp Six emerged out of the mine operation of Black Mountain, Inc. which closed in 1988 and was programmed for sale to interested parties.

References

1987 Constitution of the Philippines. (1987). The 1987 Constitution of the Republic of the Philippines. Retrieved from www.gov.ph/aboutphil/constitution.asp

Alegre, A. G. (Ed.). (1996). *Trends and Traditions, Challenges and Choices: A Strategic Study of Philippine NGOs*. Quezon City: Ateneo Center for Social Policy and Public Affairs.

Alyansa Tigil Mina. (2016, 22 August). For an Alternative Minerals Management Bill – Statement of Akbayan Senator Risa Hontiveros. Retrieved from http://alyansatigilmina.net/2016/08/22/for-an-alternative-minerals-management-bill-statement-of-akbayan-senator-risa-hontiveros/

Alyansa Tigil Mina. (2017). About Us: What Is the Alyansa Tigil Mina Organization? Retrieved from http://alyansatigilmina.net/about/

Ang See, T. (1990). *The Chinese in the Philippines: Problems and Perspectives*. Manila: Kaisa Para Sa Kaunlaran, Inc.

Bagamaspad, A., & Hamada-Pawid, Z. (1985). *A Peoples' History of Benguet Province*. Baguio: Baguio Printing & Publishing Company, Inc.

Baguio Country Club. (2015). *The Baguio Country Club, Baguio, Philippine Islands*. London: Forgotten Books.

Bello, W. (2004). *The Anti-Development State: The Political Economy of Permanent Crisis in the Philippines*. New York: Zed Books.

Bicol Today. (2012, 17 April). NASSA Supports Mining Ban in Bicol. Retrieved from http://bicoltoday.com/2012/04/17/nassa-supports-mining-ban-in-bicol/

Cabalda, M. V., Banaag, M. A., Tidalgo, P.N.T., & Garces, R. B. (2002). Sustainable Development in the Philippine Minerals Industry: A Baseline Study. Retrieved from http://pubs.iied.org/pdfs/G00614.pdf

Caballero, E. J. (1996). *Gold from the Gods: Traditional Small-Scale Miners in the Philippines*. Quezon City: Giraffe Books.

Catholic News. (2006, 18 December). Mining to Blame for Typhoon Deaths, Filipino Priest Says. Retrieved from www.cathnews.com/news/612/93.php

Catholic News. (2007, 29 January). Mining Companies Modern "Trojans", Says Filipino Bishop. Retrieved from www.cathnews.com/news/701/149.php

CBCP (Catholics Bishops Conference of the Philippines). (1998, Fabruary). A Statement of Concern on the Mining Act of 1995. Retrieved from www.philsolnl/A00a/CBCP-mining-feb98.htm

CBCP (Catholic Bishops Conference of the Philippines) News Service. (2006, 29 January). A Statement on Mining Issues and Concerns – 'Do Not Defile the Land Where You Live and Where I Dwell' (Num. 35:34). Retrieved from www.cbcpworld.com/cbcpnews/html/miningissues.html

CBCP-NASSA (Catholic Bishops Conference of the Philippines – National Secretariat for Social Action Justice and Peace). (2012, 1 October). Bishop Calls for Philex Mine Closure. Retrieved from http://nassa.org.ph/tag/bishop-calls-for-philex-mine-closure/

CBCP-NASSA (Catholic Bishops Conference of the Philippines – National Secretariat for Social Action Justice and Peace), Climate Change Congress of the Philippines, Philippine Misereor Partnership Inc.), Peace Foundation, I., PKMP (Pambansang Kaisahan ng mga Magbubukid ng Pilipinas, Katribu Indigenous Peoples' Partylist, Community Volunteer Missioners. (2012, 16–17 September). The Philex Mine Tailings Spill of 2012: An Independent Fact-finding Mission

Report. Retrieved from www.ecoteneo.org/wp-content/uploads/2012/10/cFinal-FFM-Philex-Report-2Oct-2012FINAL-REPORT.pdf

CEC (Center for Environmental Concerns – Philippines). (2011). A Primer on the People's Mining Bill (House Bill 4315). Retrieved from www.cecphils.org/system/files/CEC_Primer%20on%20the%20People%E2%80%99s%20Mining%20Bill_2011.pdf

Chaloping-March, M. (2011). *The Trail of a Mining Law: 'Resource Nationalism' in the Philippines*. Paper presented at the Conference on Mining and Mining Policy in the Pacific: History, Challenges and Perspectives, Institut de Recherche Pour Le Développement (IRD), Noumea, New Caledonia, 21–25 November.

Chaloping-March, M. (2014). The mining policy of the Philippines and 'resource nationalism' towards nation-building. *Journal of the Society of Oceanists (Journal de la Société des Océanistes)*, *138*(1), 43–56.

Chaloping-March, M. (2015a). *Gold for the Good of the Greater Lot: An Indigenous Notion of and Pursuit for Sustainability*. Paper presented at the Vulnerability and Adaptive Capacity: Proceedings of the 21st International Sustainable Development Research Society Conference, Geelong, Victoria, Australia. Retrieved from https://isdrs2015.exordo.com/files/papers/107/final_draft/2015_ISDRS_FINAL_Gold_for_the_Greater_Good_M.Chaloping-March.pdf

Chaloping-March, M. (2015b). *'Ismol Iskeyl' in Mainit: Local Control of Mineral Resources as a Postulate for Sustainability*. A Research Report Submitted to the Barangay Council of Mainit, Bontoc Mountain Province, the Mines and Geosciences Bureau – CAR, and the National Commission on Indigenous Peoples – CAR.

Clancy, M. (2005). *The Rebirth of the Philippine Mining Industry*. Makati, Philippines: Philippine Business Leaders Forum Inc.

Constantino-David, K. (1992). The Philippine experience in scaling-up. In M. Edwards & D. Hulme (Eds.), *Making a Difference: NGOs and Development in a Changing World* (pp. 137–147). London: Earthscan Publications Ltd.

DCMI (DIOPIM Committee on Mining Issues). (2006). DIOPIM Committee on Mining Issues Advocates for a Sustainable Future. *About Us*. Retrieved from http://dcmiphil.org/dcmi/about/

De Morga, A. (1609, Re-published in 1907). History of the Philippine Islands Vols 1 and 2. 10. Retrieved from www.gutenberg.org/cache/epub/7001/pg7001-images.html

DENR Administrative Order No. 2015-03. (2015, 16 March). Implementing Rules and Regulations (IRR) of Republic Act No. 7076, Otherwise Known as People's Small Scale Mining Act of 1991. Retrieved from http://server2.denr.gov.ph/uploads/rmdd/dao-2015-03.pdf

DENR Administrative Order No. 2015-07. (2015, 30 April). Mandating Mining Contractors to Secure ISO 14001 Certification. Retrieved from http://server2.denr.gov.ph/uploads/rmdd/dao-2015-07.pdf

DENR Administrative Order No. 2017-07. (2017, 10 March). Mandating Mining Contractors to Participate in the Philippine Extractive Industries Transparency Initiative (PH-EITI). Retrieved from http://server2.denr.gov.ph/uploads/rmdd/dao-2017-07.pdf

DENR Administrative Order No.1996-40. (1996, 19 December). The Revised Implementing Rules and Regulations of Republic Act No. 7942, Otherwise Known as the "Philippine Mining Act of 1995". Retrieved from www.mgb.gov.ph/images/stories/MIN_DAO_1996-40.pdf

DENR (Department of Environment and Natural Resources). (2006). DENR Assessment of the Rapu-rapu Polymetallic Project. Retrieved from www.mgb.gov.ph/miningissues/rapurapu/rapurapuassessment.pdf

Disini, A. F. (1999, December). Regional review: The Philippines. *Mining Annual Review*, A175.

Domingo, E. G. (1993). The Philippine mining industry: Status and trends in mineral resource development. *Journal of Southeast Asian Earth Sciences*, *8*(1–4), 25–36.

Domingo, E. G. (2003, 8–9 October). *Public Policy Initiatives in Institutionalising CSR in the Philippine Mining Industry*. Paper presented at the World Bank International Conference on Public Policy for Corporate Social Responsibility, Washington, DC.

The Economist. (1983). Marcos is short of cash to bail out his chums. No. 7286.

Executive Order 79 (2012, 6 July). Institutionalizing and Implementing Reforms in the Philippine Mining Sector Providing Policies and Guidelines to Ensure Environmental Protection and Responsible Mining in the Utilization of Mineral Resources. Retrieved from http://denr.gov.ph/images/stories/Executive_Order_79_2012.pdf

Executive Order No. 192 (1987, 10 June). Providing for the Reorganization of the Department of Environment, Energy and Natural Resources Renaming it as the Department of Environment and Natural Resources. Retrieved from http://forestry.denr.gov.ph/pdf/eo192s.1987.pdf

Executive Order No. 270. (2004, 16 January). National Policy Agenda on Revitalizing Mining in the Philippines. Retrieved from www.mgb.gov.ph/policies_eo270.htm

Executive Order No. 279. (1987, 25 July). Authorizing the Secretary of Environment and Natural Resources to Negotiate and Conclude Joint Venture, Co-Production, or Production-Sharing Agreements for the Exploration, Development and Utilization of Mineral Resources, and Prescribing the Guidelines for Such Agreements and Those Agreements Involving Technical or Financial Assistance by Foreign-Owned Corporations for Large-Scale Exploration, Development, and Utilization of Minerals. Retrieved from www.lawphil.net/executive/execord/eo1987/eo_279_1987.html

Gonzalez, J. L., III. (2001). Philippines: Continuing people power. In J. Funston (Ed.), *Government and Politics in Southeast Asia* (pp. 252–290). Singapore: Institute of Southeast Asian Studies.

Hilhorst, D. (2003). *The Real World of NGOs: Discourses, Diversity and Development*. London and New York: Zed Books.

House Resolution No. 600 (2016, 5 December). Resolution Expressing the Sense of the House of Representatives Condemning the Lifting of Moratorium on Large Scale Mining. Activities in the Province of Capiz. Retrieved from http://www.congress.gov.ph/legisdocs/basic_17/HR00600.pdf

Howard, M. C. (1988). *The Impact of the International Mining Industry on Native Peoples*. Sydney: Transnational Corporations Research Project, University of Sydney.

Human Rights Online Philippines. (2015, 15 January). CBCP-NASSA and ATM Joint Statement on the Mining Issue During the Visit of Pope Francis Manila and Tacloban in the Philippines: 15–19 January 2015. Retrieved from https://hronlineph.com/2015/01/16/statement-cbcp-nassa-and-atm-joint-statement-on-the-mining-issue-during-the-visit-of-pope-francis-manila-and-tacloban-in-the-philippines-15-19-january-2015/

Keesing, F. M., & Keesing, M. (1934). *Taming Philippine Headhunters: A Study of Government and of Cultural Change in Northern Luzon*. London: Allen & Unwin.

Leonen, M. F. (2004). Weaving worldviews: Implications of constitutional challenges to the indigenous peoples rights act of 1997. *Social Policy and the Law (Journal of the Integrated Bar of the Philippines), 30 1st and 2nd Quarters 2004*(1), 153–184.

Lopez, S. P. (1992). *Isles of Gold: A History of Mining in the Philippines*. Singapore: Oxford University Press.

Lynch, O. J. (1986). Philippine law and upland tenure. In S. Fujisaka, P. Sajise, & R. Castillo (Eds.), *Man, Agriculture, and the Tropical Forest: Change and Development in the Philippine Uplands* (pp. 269–292). Bangkok: Winrock International Institute for Agricultural Development.

Lynch, O. J., & Talbott, K. (1995). *Community-Based Forest Management and National Law in Asia and the Pacific*. Washington, DC: World Resources Institute.

Malig, M. L. (2002). Sustaining Poverty: The National Report on Sustainable Development in the Philippines. *Focus on the Global South*. Retrieved from www.focusweb.org/publications/Books/National-report-on-sd.pdf

The Manila Times. (2006, 6 February). Bishops' antimining drive breathes life to old debate. Retrieved from www.manilatimes.net/national/2006/feb/06/yehey/top_stories/20060206top5.html

Mendoza, H. D. (2014, 27 June). Eco-Friendly Recovery of Gold and Copper for the Small-Scale Mining Industry. Retrieved from www.pcieerd.dost.gov.ph/images/downloads/presentation_materials/pcieerd4thanniversary/session_a/1_UP-DOST_Presentation_27_%20June_2014_copy.pdf

Minda News. (2011a, 11 June). South Cot Approves Environment Code, Bans Open Pit Mining. Retrieved from www.mindanews.com/environment/2010/06/11/south-cot-approves-environmentcode-bans-open-pit-mining/

Minda News. (2011b, 18 February). Tampakan Project is 'Catastrophe in the Making', Militant Group Says. Retrieved from www.mindanews.com/topstories/2011/02/18/tampakan-project-is-%E2%80%98catastrophe-in-themaking%E2%80%99-militant-group-says/

Mines and Geosciences Bureau. (2000). *A Response to the Issues Raised against Mining*. Diliman, Quezon City: Mines and Geosciences Bureau, Department of Environment and Natural Resources.

Mines and Geosciences Bureau. (2004). Highlights of the Small-Scale Mining Law and Its Implementing Rules and Regulations. Retrieved from www.mgb.gov.ph/media%briefing_highlights-ssm.htm

Mines and Geosciences Bureau. (2005). Rapu-Rapu Mines Pour First Gold. Retrieved from www.mgb.gov.ph/news/072005_rapurapu.htm

Mines and Geosciences Bureau. (2017a, 21 April). Mining Industry Statistics. Retrieved from http://mgb.gov.ph/attachments/article/162/MIS(2015)%20(1)%20(1).pdf

Mines and Geosciences Bureau. (2017b, 5 April). Philippine Metallic Mineral Production. Retrieved from www.mgb.gov.ph/images/Metallic-Mineral-Production-2016-vs-2015-for-mgb-website-05-Aprll-2017.pdf

Mines and Geosciences Bureau. (2017c, January). The Philippines Minerals Industry at a Glance. Retrieved from www.mgb.gov.ph/images/homepage-images/mining-facts-and-figures-updated-January-2017.pdf

The Mining Journal. (1995, 18 August). Philippines' road to recovery, 118.

Miralao, V. A., & Bautista, M.C.B. (1993). The growth and changing roles of NGOs and the voluntary sector. *Philippine Sociological Review*, *41*(1–4), 19–36.

MISN (Mining Injustice Solidarity Network). (2010). PCSD Endorsement to Macroasia Multi-Billion Giant Deferred: An Initial Victory for NGOs and Indigenous Peoples. Retrieved from www.solidarityresponse.net/palawan-unesco-man-and-biosphere-reserve-sold-out-to-mining-companies-last-government-decision-on-30-july/

NCIP Administrative Order No. 2012-3 (2012, 13 April). The Revised Guidelines on Free and Prior Informed Consent (FPIC) and Related Processes of 2012. Retrieved from http://ncipcar.ph/pdf_files/admin_orders/ncip%20ao%20no%203,%20s%20 2012%20(fpic).pdf

NSCB (National Statistics Coordination Board). (2015). Provincial Summary Number of Provinces, Cities, Municipalities and Barangays, by Region, as of June 2015. Retrieved from http://nap.psa.gov.ph/activestats/psgc/NSCB_PSGC_SUMMARY_June2015.pdf

Pasimio, H. J. (2005, November). Extracting Growth from Mining. Retrieved from www.senate.gov.ph/publications/PI%202005-11%20ExtractingGrowth%20from%20Mining.pdf

Ph-EITI (Philippines Environmental Transparency Initiative). (2016, December). Extracting Value in Transparency: The Third PH-EITI Report (FY2014). Retrieved from https://eiti.org/sites/default/files/documents/ph-eiti_country-report_2016_1.pdf

Phelan, J. L. (1959). *The Hispanization of the Philippines Spanish Aims and Filipino Responses, 1565–1700*. Madison: University of Wisconsin Press.

Philippine Bill of 1902. (1902, 1 July). An Act Temporarily to Provide for the Administration of the Affairs of Civil Government in the Philippine Islands, and for Other Purposes. Retrieved from www.gov.ph/constitutions/the-philippine-organic-act-of-1902/

Plumlee, G. S., Morton, R. A., Boyle, T. P., Medlin, J. H., & Centeno, J. A. (2000). An Overview of Mining-Related Environmental and Human Health Issues, Marinduque, Island, Philippines: Observations from a Joint U.S. Geological Survey – Armed Forces Institute of Pathology Reconnaissance Field Evaluation, May 12–19, 2000. Retrieved from https://pubs.usgs.gov/of/2000/ofr-00-0397/ofr-00-0397.pdf

Presidential Decree No. 463. (1974, 17 May). Providing for a Modernized System of Administration and Disposition of Mineral Lands and to Promote and Encourage the Development and Exploitation Thereof – *Full Text Published Online by the*

Chan Robles Virtual Law Library. Retrieved from www.chanrobles.com/presidentialdecrees/presidentialdecreeno463.html

Prill-Brett, J. (1994). Indigenous land rights and legal pluralism among Philippine highlanders. *Law & Society Review*, *28*(3, Law and Society in Southeast Asia), 687–698.

Prill-Brett, J. (2003, 11–14 July). *Changes in Indigenous Common Property Regimes and Development Policies in the Northern Philippines*. Paper presented at the RCSD International Conference Politics of the Commons: Articulating Development and Strengthening Local Practices, Lotus Pang Suan Kaew Hotel, Chiang Mai, Thailand.

Prill-Brett, J., & Salinas-Ramos, T. (1998). *Tuba Ethnohistory*. Baguio City: Cordillera Studies Center, University of the Philippines College Baguio.

Ramos, F. V. (2005). *The Social Responsibility of the Mining Industry*. Paper Keynote Speech at the International Mining Investment Conference, organized by the Chamber of Mines of the Philippines. New World Renaissance Hotel, Makati City, Philippines. 3 February

Ramos, H. C., Cabalda, M. V., & Banaag, M. A. (2000). *Tailings Dam Accidents and the Use of Chemicals in Mining: Issues, Policy Response and Lessons Learned from the Philippines*. Paper presented at the UNEP/Government of Australia International Workshop on Environmental Regulation for Accident Prevention in Mining – Tailings and Chemicals Management, Perth, Australia.

RFFC (Rapu-Rapu Fact-Finding Commission). (2006). Findings and Recommendations of the Fact-Finding Commission on the Mining Operations in Rapu-Rapu Island. Retrieved from www.mgb.gov.ph/miningissues/rapurapu/rrffcreport.pdf

Republic Act No. 6364. (1971, 5 August). An Act Amending Republic Act Numbered Thirty Hundred Eighty-Nine, as Amended, Entitled 'An Act to Provide for Assistance to the Gold Mining Industry'. Retrieved from http://laws.chanrobles.com/republicacts/64_republicacts.php?id=6368

Republic Act No. 7160. (1991, 10 October). *An Act Providing for a Local Government Code of 1991*. Retrieved from http://www.lawphil.net/statutes/repacts/ra1991/ra_7160_1991.html.

Republic Act No. 7942. (1995, 10 February). An Act Instituting A New System of Mineral Resources Exploration, Development, Utilization and Conservation. Retrieved from http://www.mgb.gov.ph/images/stories/RA_7942.pdf

Republic Act No. 8371 (1997, 29 October). An Act to Recognize, Protect and Promote the Rights of Indigenous Cultural Communities/Indigenous People, Creating a National Commission of Indigenous People, Establishing Implementing Mechanisms, Appropriating funds Therefor, and for Other Purposes." Retrieved from http://www.officialgazette.gov.ph/1997/10/29/republic-act-no-8371/

Republic Act No. 7941. (1995, 3 March). An Act Providing for the Election of Party-List Representatives Through the Party-List System, and Appropriating Funds Therefor. Retrieved from www.comelec.gov.ph/?r=References/RelatedLaws/ElectionLaws/RA7941

RFFC (Rapu-rapu Fact Finding Commission Report). (2006). Findings and Recommendations of the Fact-Finding Commission on the Mining Operations in Rapu-Rapu Island. Retrieved from www.mgb.gov.ph/miningissues/rapurapu/rrffcreport.pdf

Scott, W. H. (1970). Igorot responses to Spanish aims: 1576–1896. *Philippine Studies, 18*(4), 695–717.

Scott, W. H. (1974). *The Discovery of the Igorots: Spanish Contacts with the Pagans of Northern Luzon*. Quezon City: New Day Publishers.

Senate Bill No. 1069. (2016, 25 August). An Act to Regulate the Rational Exploration, Development, and Utilization of Mineral Resources, and to Ensure the Equitable Sharing of Benefits for the State, Indigenous Peoples and Local Communities, and for Other Purposes. Retrieved from http://senate.gov.ph/lisdata/2465921252!.pdf

Serrano, I. R. (2002). NGOs and NGO-GO Relations in the Philippines. Retrieved from www1.biz.biglobe.ne.jp/~iam/httpdocs/delhi02/paper3.html

Silliman, G. S., & Garner Noble, L. (Eds.). (1998). *Organizing for Democracy: NGOs, Civil Society, and the Philippine State*. Honolulu: University of Hawaii Press.

Supreme Court of the Philippines. (1995, 16 January). G.R. No. 98332. Miners Association of the Philippines, Inc., Petitioner, Vs. Hon. Fulgencio S. Factoran, Jr., Secretary of Environment and Natural Resources, and Joel D. Muyco, Director of Mines and Geosciences Bureau, Respondents. Retrieved from www.lawphil.net/judjuris/juri1995/jan1995/gr_98332_1995.html

Supreme Court of the Philippines. (2000, 6 December). G.R. No. 135385: Cruz vs Sec. of Environment and Natural Resources. Retrieved from www.supremecourt.gov.ph/jurisprudence/2000/dec2000/135385.htm

Supreme Court of the Philippines. (2004, 24 January). G.R. No. 127882: La Bugal-B'laan Tribal Association Vs. Ramos. Retrieved from http://sc.judiciary.gov.ph/jurisprudence/2004/jan2004/127882.htm

Symon, A. (2007). Petroleum and mining in Southeast Asia: Managing the environmental and social impacts. *Southeast Asian Affairs*, 77–100.

Tan, A.K.J. (1998, 15 September). Preliminary Assessment of Philippines' Environmental Law (APCEL Report: Philippines). Retrieved from http://law.nus.edu.sg/apcel/dbase/filipino/reportp.html

Walker, S. (1992, March). Philippine mining. *Mining Magazine*, 136.

Wurfel, D. (2004). Civil society and democratization in the Philippines. In Y. Sato (Ed.), *Growth and Governance in Asia* (pp. 215–223). Honolulu, Hawaii: Asia-Pacific Center for Security Studies.

Zuniga, J. D. (1966). *An Historical View of the Philippine Islands*. Manila: Filipiniana Book Guild.

3 Pre-closure challenges at the Padcal mine

> The company has said again that the mine will close one day. Sometime in the early 1990s, I heard that the company is going to close in 1997. But look, we are still here. It is 2005 now, and we are told that the mine is closing in 2011. I think the company will be able to find ores to dig, and we can continue to live our happy lives. Closure of the mine will not happen. No, I don't think so.[1]
>
> Ildefonso Rempillo (pseudonym)
> A local official of Barangay Camp Three

After it began operations in 1958, Philex Mining Corporation (PMC) has had a colourful journey: growth in the 1960s towards its glory days in the 1970s and 1980s, deficits during most of the 1990s, but emerging as one of only two[2] survivors among the country's scores of medium- to large-scale mines towards the third millennium. Other large-scale mines have not been as resilient to withstand the period of grimly low metal prices and high production costs that characterised the industry during most of the 1990s. The third millennium has offered more gruelling challenges to PMC. Nonetheless, its president who is at the same time the chief executive officer (CEO) and chief operating officer (COO), referring to a catastrophe that he considers the worst incident in the entire company history, notes: 'For like gold, fire does not destroy us. Instead, our malleability survives this process of change. When everything is forged and cast, we reemerge, core intact, bearing a luster that will never dull' (PMC, 2013b).

PMC has built its huge occupational community in Padcal. Philex is the minesite camp which takes its name off the company's. It is comprised of PMC's mine workers and their families who regard themselves well-nurtured by the company. Reciprocally, PMC has enormously benefited from its generally stable workforce and the support of workers' families. The company exemplifies the truism that employees are the key to successful business operations. Without the persevering mass of mine workers, the company would not have achieved its corporate objective of generating and maximising profits. The provision of infrastructure, housing and basic amenities for workers and their families is essentially for facilitating the efficient operation of the mine and maintaining manpower stability and productivity. Correspondingly, residents of Philex have clung to the company throughout many decades to a point that the idea of departing from the minesite camp is, as what would clearly emerge, a reality that took long to permeate in the consciousness of many.

Weathering the huge storms

In the late 1980s, PMC's exploration strategy centred on finding gold reserves, which resulted in the company's acquiring the Bulawan gold mine in Negros Occidental in Central Philippines. The Bulawan mine began operating in January 1996 but was closed in 2002 and placed under care and maintenance due to low metal prices. PMC also acquired in February 2009 the Silangan project in Surigao del Norte in Northern Mindanao, which involves developing gold, copper, and silver deposits (PMC, 2012a). As the closure of the Padcal mine approaches, the Silangan project favourably beckons, although shifts in policy and changes may avert positive prospects.

As PMC moved into the third millennium, the Philippine Security Exchange Commission approved on 23 June 2004 the extension of the company's corporate life for another 50 years (PMC, 2007, p. 47), notwithstanding the depletion of the ores of the Santo Tomas II deposit which signalled the impending closure of the Padcal mine. The company started informing the residents of Philex and those in the surrounding communities about the looming closure of the mine. The year 2011 was projected as the closure date. In 2005, PMC implemented a scheme to literally decongest its huge occupational camp of more than 12,000 people.

As the eventual closure became more perceptible, the year 2011 saw the highest prices for copper and gold in the previous 11 years. Ironically, 2011 was the year the Padcal mine was projected to close (but updated later). As though the metal prices during the year were not propitious enough for PMC, the Padcal mine's ore tonnage was the highest since 1987, surpassing the year's target tonnage (PMC, 2012b). However, countering the auspicious prospects were socio-political challenges such as the increasing scrutiny of the mining industry by well-informed and firmly-networked environment and nationalist organisations, whose opposition to mining – as the former DENR Secretary Elisea Gozun described in 2003 – did not emerge from sheer imagination. Anti-mining groups have revelled in the move of the national government in 2011 to review the current fiscal regime for mining. With the help of party-list organisations, mining critics have been pushing for the passage of an alternative mining law to replace the Mining Act of 1995. In addition, moratoriums against mining have been imposed by several local government units across the country. For the year 2011, a few provinces passed legislation disallowing large-scale mining within their jurisdictions: Albay (effective March 2011), Bukidnon (effective May 2011, reinforcing a related ordinance passed in 2009), and Zamboanga del Norte banning specifically open pit mining (effective November 2011).

Against these multiple threats that have not shown any abatement, PMC strategized to strengthen itself to brace the storms waging against the mining industry. PMC's senior management has undergone a series of strategic planning exercises, and redefined the company's new vision: 'to be a highly respected world-class natural resource company committed to deliver excellent value to its investors, employees and other stakeholders' (PMC, 2012b, p. 4).

Ironically, on 1 August 2012, PMC found itself facing what turned out to be an immobilising disaster. The company's third tailings storage facility (TSF 3) released sediments through one of the dam's penstocks (PMC, 2013b). It was a cataclysmic spillage which national and local media have covered extensively. It raised questions about PMC's status of having an environmental management system that is ISO 14001-certified, and being a consistent recipient of distinction awards in environmental management. The TSF event generated condemnation from environment organisations and anti-mining groups. For instance, Environment Justice Atlas (2015) has labelled the Padcal mine spillage as the 'biggest mining disaster in the Philippines.' The Mines and Geosciences Bureau (MGB) ordered operations at the Padcal mine to halt until the company ensure the safety and integrity of its TSF. The Pollution Adjudication Board imposed on 18 January 2013 a hefty PhP92.8 million[3] fine on PMC for violating the Philippine Clean Water Act of 2004 and the terms of its environmental compliance certificate. In February 2013, PMC also paid the national government PhP1.03 billion[4] to cover environmental fees for the costs of remediation and rehabilitation activities of the affected area.

PMC tried to assuage the fears of the affected communities along the Balog Creek and downstream towards the Agno River by assuring that the sediment discharge was nontoxic based on results of independent studies[5] of scientists and academicians (PMC, 2013a). In November 2012, the tireless employees of PMC, after three months of working day and night, eventually stopped the discharge of tailings by blocking the sinkhole in one of the two penstocks, and sealing the underground tunnel of the affected penstock (PMC, 2013b). This has allowed the company employees to begin carrying out the required remediation and rehabilitation program along a

2.8-kilometre stretch of the Balog Creek up to its convergence with the Agno River (Terso, 2017). In February 2013, the MGB allowed the Padcal mine to operate for four months to undertake further remediation measures. Finally, on 27 August 2014, the MGB lifted its suspension order, and PMC resumed operations (PMC, 2015, p. 12). This action by MGB drew anger and criticism from militant groups and environmental activists (Kalikasan-PNE, 2013)

Among Philex residents, their close-knit relationships and strong sense of belonging – to one community and one company – was tested when they all had to deal with the tailings spillage that brought the company to a standstill, and posed a disturbingly uncertain future. At present, residents look back to the event which they consider as an inordinately distressing time, but one that they all shared and won over. They recall the ominous mood in Padcal and how uncertainty had loomed as it took two years before the MGB gave permission for PMC to resume operations. Some residents consider the tailings spillage as a 'test of fire' for both the company and the community (Terso, 2017). All the employees, regardless of rank and role or type of work in the company, age or physique, under tense and uncommonly draining conditions, gave their share of strain for the required remediation and rehabilitation program by reporting daily for work 'down by the river' – in Balog Creek and its confluence with the Agno River. Some members of worker families joined the employees in the clean-up and rehabilitation. They were happy for the opportunity to help but receiving income for it was an additional reward. Some groups of residents together with school-children, teachers, and villagers in Ampucao and Camp Three travelled to Manila and called on MGB to lift the suspension on PMC.

PMC celebrated its 60th year in December 2016. Using a grade school boy's poster in a contest depicting 'a man wearing a hard hat and a curly-haired man in a red shirt linked by a chain of hands', PMC conveys in its company webpage the message, 'Philex the community is the sum total of what the company has become, and not the mining firm that started in Benguet'(Philippine Daily Inquirer, 2016).

A brief history of Philex Mining Corporation

PMC began with Mr. Henry A. Brimo who, following the Second World War, was recruited into the Baguio Gold Mining Company, an old mine adjacent to Antamok in Itogon. Incorporated in 1930 and made operational by a small group of American soldiers-turned-miners, Baguio Gold was among the viable mines during the pre-war years. Mr. Brimo became Baguio Gold's president in 1947 and had since worked in the mining industry until his retirement in 2002. While the small team of Americans effectively managed the affairs of Baguio Gold, it was Mr. Brimo who directed the rehabilitation of the company during the postwar period (Lopez, 1992, p. 86). Baguio Gold was the predecessor of PMC with Mr. Brimo as founder, president and director, then chairman, all for over 50 years.

Incorporated in the Philippines in 1955, PMC began mining operations out of the deposit of the Santo Tomas II ore body in Padcal. The company's main product is copper, with gold and silver as by-products. Production started in 1958 with a modest volume of 1,500–2,000 tonnes per day and an initial reserve of 18 million metric tonnes (PMC, 1958, p. 5). Following a series of exploration work and successive mine and mill expansions, the company's production tonnage expanded in the 1970s to 28,000 tonnes per day. During the period 1958—1959, the company produced copper by open pit method. More than four decades later, the open pit would unleash land compensation issues that would dramatically shape the company's relations with the *sitios* (sub-villages) surrounding the mine.

In 1960, the company shifted to underground block caving using the slusher system. Three decades later, it moved into a more mechanised load-haul-dump (LHD) equipment (PMC, 1995). This made the Padcal copper mine the first block cave operation in the Far East (PMC, 2007). PMC's mechanisation and modernisation in ore extraction would soon have significant bearing on its workforce size. This, in turn, would have important implications for the nature of social and economic concerns that confront the company as mine closure became imminent by the late 1990s. By this time, the company's huge workforce and their families already grew into a self-contained community (see Image 3.1) with a population of over 20,000.

From 2000 onwards, the company has continued upgrading its mine and mill equipment to further operate more efficiently. This move has been simultaneous with the development of the 782 Meter Level, the mine's last section for ore reserves. Officially commissioned in January 2006, the 782 Meter Level has ores with higher-than-average grade which gave the company some optimism that the mine will carry on operating at least up to 2011 and, if metal prices and operating costs play well, as they did, would allow the company a few more years to carry on (PMC, 2006).

Beginning 2005, PMC has achieved record-high revenues. The year 2005 marked the company's 50th year of existence and began hopes for a return to the company's apex days of profits. The continuing rise of metal prices in the world market, coupled with an increase in the grade of ore milled, resulted in extremely favourable profits. The company's consolidated net income in 2005 was a lucrative PhP409.1 million, compared with net losses of PhP15.8 million in 2004 and PhP481.3 million in 2003 (PMC, 2006, p. 7). However, the subsequent years were much more propitious: PP3.087 billion in 2006 and an unprecedented PhP5.005 billion in 2007 (PMC, 2008, p. 9).

Image 3.1 Partial view of Philex, the minesite camp in Padcal, Tuba, Benguet, August 2016

Source: Author's research file

While metal prices declined in the last quarter of 2006, the increases in the company's gold and copper production levels, owing to better grade of ore, coupled with recovering prices of both copper and gold during the first quarter of 2007, placed PMC in a position envied by many companies. Beginning 2008, PMC's consolidated net income declined to PhP2.800 and stagnated at PhP2.735 billion in 2009 (PMC, 2010, p. 22). Despite the tailings spillage in 2012 and the suspension of the Padcal mine operations until the suspension was lifted in August 2014, PMC has continued staying profitable with consolidated net incomes of PhP312 million in 2013 (despite net loss of PhP311 million) in 2012), PhP703 million in 2014, PhP 776 million in 2015, and PhP1.567 billion in 2016 (PMC, 2017, p. 5).

Philex: the minesite camp in Padcal

Padcal is the site of the copper mine of PMC. It is found within Tuba which, like Itogon, lies at the southern part of Benguet Province. Padcal is a ridge whose terrain is characterised by steep slopes and deep gullies. One of my Ibaloy resource persons explains that Padcal was so named after the characterisation of the area that was '*Pakshal*', short for '*Nai-pakshal*', a descriptive Ibaloy term denoting ‘outstretched’ or ‘protruding’. Padcal embodies a spur projecting eastward from the main mass of the mountain. Padcal was an uninhabited hunting ground for wild boars and deer until the early 1950s. Large sections surrounding the area used to be grazing land for cattle. In 1957, PMC developed the mine and built the mill (PMC, 1958). With mine production eventually taking off in 1958, migrants soon arrived in Padcal as PMC recruited workers. Subsequently, the company built several housing facilities towards the 1960s. In the 1970s, more structures had been put up such as the elementary and secondary schools, hospital, basketball courts, stores, a bowling alley, and offices. Thus, a virtually self-contained boomtown, named Philex, emerged in Padcal.

The idea of a minesite camp composed mainly of people from different provinces, coming together in one geographic area, was not new in Benguet. At the time Philex was carrying out its exploration in Padcal in the 1950s, large mining settlements had already been established by Benguet Corporation in neighbouring Antamok, Acupan and Balatoc. Similarly, in the northern part of the province was Lepanto, a large community that developed through the operation of Lepanto Consolidated Mining Company, established in 1936 from the revival of a Spanish-era company.

By political-administrative jurisdiction, the Padcal mine and its occupational community, Philex, straddle two barangays (see Figure 3.1) – Camp Three, which is part of Tuba municipality; and Ampucao, which is part of Itogon municipality. Thus, PMC deals with two groups of local government officials who sometimes disagree about permitting approvals, and splitting tax revenues and social development projects.

The ethnolinguistic groups in both Tuba and Itogon are primarily the Ibaloy. Philex’s population was highest in the 1980s when the company had over 5,000 employees, making the total population 22,000. Due to the slump of metal prices in the following few years, the company had its first retrenchment in 1993, thereby reducing the population to about 18,000. With a workforce of 2,698 in 2005, the community’s population stood at about 12,000 and has continued dwindling. In 2012, the Padcal mine had 2,123 employees. By August 2016, the number of employees was 1,879.

Residents of Philex regard the minesite camp as secure and safe for their families, particularly the schoolchildren. They proudly talk about the curfew at night, the absence of traffic, and the proximity of workplace to their dwelling units. Some residents get to learn to speak other dialects even without trying as they mingle with people coming from different provinces. Over several decades – at least until 1993 when the first retrenchment of workers took place – residents have all been living together in one place, calling themselves *Taga-Philex* (‘of / from Philex’), and even coining the term *Philexian*.

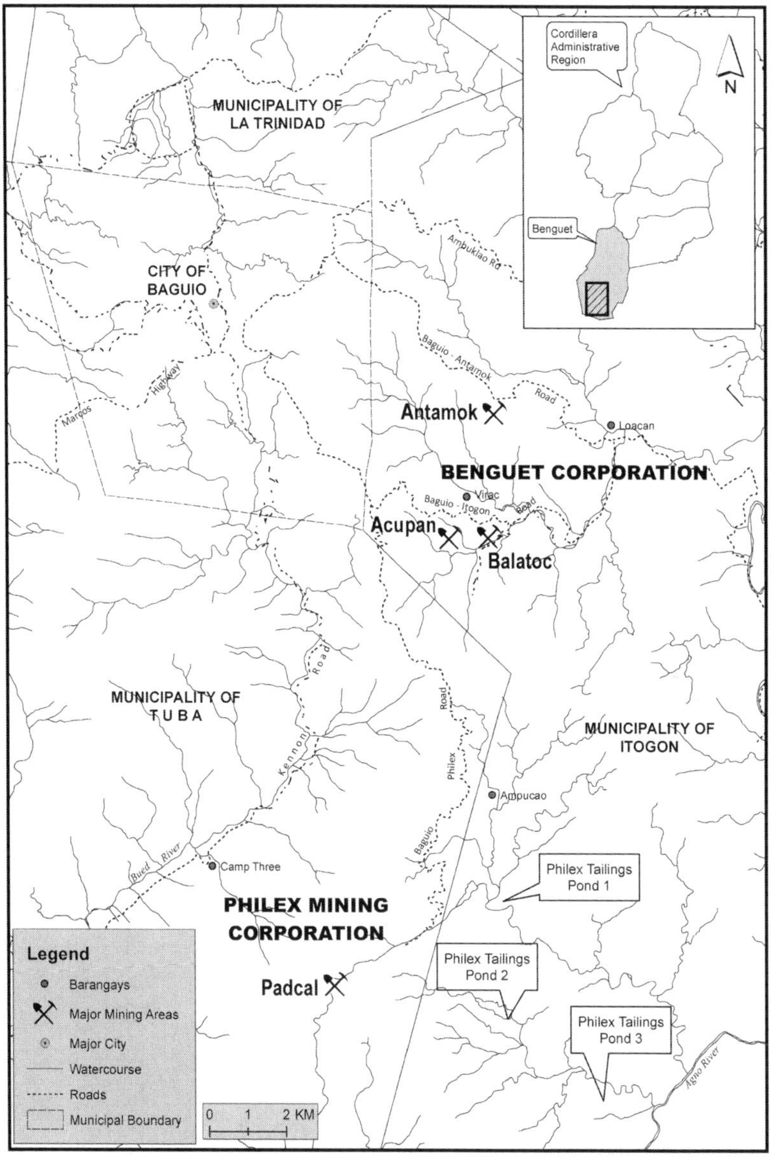

Figure 3.1 The tenement area of Philex Mining Corporation and its minesite camp in Padcal showing location within two barangays of two adjoining municipalities

Perceived favour and bounty

Philex residents have perceived themselves fortunate for their employment, privileges, and entitlements. This general perception can be understood in the light of what many Filipinos families need: stable income, secure jobs, and capability to send their children to school. Under the leadership of the company's founder and president, the late Henry A. Brimo,[6] PMC has succeeded in laying down one lasting thought that took root deeply as a sentiment shared by residents: the people of Philex are one big family. Mr. Brimo's paternal approach of running the company and dealing with employees and other residents in the minesite camp is well described in the narratives of older workers and retirees. The remark *Kasla annak ti arig ni Apo Brimo kadakami* ('Mr. Brimo treated us as his children') has been oft-cited in my conversations with residents, even without my asking about the company president. Employees of different ranks and roles talk about Mr. Brimo's persona as something that made PMC the company they have been happy and loyal to work for: generous, understanding and exceptionally approachable. Many residents recall how Mr. Brimo would travel from Manila to Philex every weekend and join camp residents in attending mass at the St. Joseph Catholic Church located within Philex itself. Others talk about Mr. Brimo's spending his birthday in Philex every year, and workers and their families join in the celebration. The company president has been a corporeal figure in Philex, not just managing the company but also walking in the streets and conversing with residents.

Mr. Brimo's message in 1964, following the negotiations with the workers' union for further wage increases, offers a glimpse of how he regarded the PMC employees:

> It is company policy to afford its workers reasonable and satisfactory pay rates, including the usual, and sometimes by necessity, unusual fringe benefits. In doing so, it acts on the principle that a satisfied labour force is a good investment, even though the cost may appear high.
>
> PMC (1965, p. 3)

To some extent, the affinity of Philex residents for the company blurs the demarcations between what is personal/family and work/company. The obscurity is demonstrated in the way residents have used company facilities for their own family and personal purposes. Ultimately, it is also displayed in the reluctance, sometimes refusal or denial, of many residents to accept that the mine in Padcal is going to close. The profound emotional-psychological and economic impacts of mine closure on many workers and their families cannot be in any way fully gauged.

Most PMC employees, particularly the 'first generation workers' – those employed starting the 1950s and 1960s whose grown-up children are also now working, with many of them at PMC – are grateful for having been employed by the company. Many employees, both older and younger, including the current president and CEO and COO, have never worked for any other employer during their entire working lives. They talk about friendships for life that have been built around relationships as co-workers and as co-Philexians.

Older residents describe the yearly community *fiestas* (festival) during which the company invites an elder from among the sitios around Padcal to grace the *cañao.*[7] The native 'priest' would offer thanksgiving for the health and safety of all workers and the company for meeting its production goals. Philex would be swamped by visitors coming from surrounding villages. Residents talk of wage increases each year, and the multiple bonuses they receive including premiums when company production targets are attained. The vice president of the Philex Miners' Union (PMU), Local NAMAWU 101, explains that the minimum wage for Philex employees is 62 percent higher than the prescribed government rate in the Cordillera Administrative Region and 10 percent more than that of the National Capital Region. As early as 1961 when PMC signed the first three-year bargaining agreement with the PMU, the company granted the workers' demand for periodic wage increases with each passing year. In this very first bargaining agreement, the company advanced the dates of effectivity of the second and third wage increases 'in view of the rather unexpected improvement in operating results, commencing the year following the signing of the agreement' (PMC, 1963, p. 7). The Philex Supervisors and Classified Employees' Union (now Philex Mines Supervisors Employees Union) members have likewise been granted pay increases. The same increases were applied to the employees who were not members of either of the two unions. Workers explain that the cordial relationship between management and their union is a result of regular meetings and prompt resolution of problems. The remarks of Rolando, a 54-year old worker at the Mill Department, provides an idea of the worker-management dynamics:

> Here in Philex, it is different. Apo[8] Brimo talks to lowly people like ourselves. He pays us attention during the meetings, and reminds us that the company is not limitless. We workers are not unreasonable because we understand what our president says. We consider the company as a hen laying eggs. We ask for the eggs, and not for the head, as is the case of other labour unions. Of course, we help the hen to stay productive. We are unashamed to tell Mr. Brimo what we want; he tells us what the company can give and cannot give and explains us the circumstances. He is like our father. We listen to him because he listens to us.[9]

The general feeling of being 'fortunate' tends to self-infect every PMC worker and Philex resident, reinforced by the felt support of no less than a much-respected and endeared company president. PMC employees have salary bonuses, which some housewives described as *nawadwad* (ample). In addition to the 13th-month salary which is mandated by law, PMC employees had received mid-year bonus (every June, the start of the school year) to help parents pay tuition fees of children at university, 14th-month bonus, and *extra-gratia* ('out of goodwill') beginning the 1980s.[10] Many residents talk about 'floating money' in the 1980s. Some housewives recall how several trucks from Baguio were frequently delivering appliances for many families in Philex who were in a purchasing frenzy. The extraordinarily high prices for copper and silver in 1980, especially during the first quarter, brought the company the highest gross revenue and net profit. The favourable prices during the early part of the decade were the result of a combination of developments including political (Iran hostage crisis), social (U.S. copper industry strike), and economic (inflation and interest rates) factors (PMC, 1981b, p. 3). On the company's 25th Anniversary in June 1980, part of the celebration included distributing cash gifts to employees, with the total amounting to over PhP1 million (PMC, 1981b, p. 4).

Given the comparatively high income of workers in Philex, it is not surprising that as early as 1980, the Vice President for Operations wrote a message as an admonition to the residents:

> From inside our little Philex world we may feel discreetly different, superior and even slightly disconnected from the rest of the Philippines. Indeed, this is a false and eventually insecure attitude because outside influences are much stronger and more durable than our own. If peace and order deteriorates elsewhere that condition would soon become our problem. Rampant inflation in the Philippines and worldwide immediately adversely affects the Philex labour-intensive and high energy-consuming operation. Added to this looms the unassailable fact that a mine is a wasting asset wherein, measured by the vast expanse of time, yesterday we found the ore and developed it, today we mined it and enjoyed the proceeds and tomorrow we awake to find it has vanished by the relentless process of depletion.
>
> PMC (1980b, p. 2)

The message is part of an editorial of the Philex Cu-Au Bulletin, the quarterly community publication which serves as Philex's primary source of news about company operations and community activities. While most PMC employees felt more endeared to the company for the fringe benefits they receive, a few individuals, such as the wife of one employee who already left the company, has her own impressions:

> My husband was retrenched in 1999. To be employed with a mining company and reside in a minesite camp is a big blessing. But it is also a reality that within the company, there is a lot of nepotism and politics. To survive, you must be good at relating with those in the higher ranks, otherwise promotion does not come to you. My husband had been bypassed a few times in promotions simply because he was not his boss's favourite. When I learned that my husband was one of those who were to be retrenched in 1999, I went to see his boss one evening and pleaded to him to consider my husband's good performance at work, and that we have children who are still studying in the elementary and high school. My husband was retrenched anyway. I am hurt because the company was not fair. My husband's termination from work led me to take up this occupation – vending vegetables in the market. This has allowed us to continue residing within the camp and see my children continue their schooling, although there is a deadline when we must leave.[11]

The Cordillera Peoples Alliance (CPA), a Baguio-based NGO, also criticised PMC for adopting labour contractualisation, a practice that involves 'hiring workers on short-term successive contracts of employment with or without gaps and then re-hiring these same workers to perform essentially the same tasks for which they were initially hired' (BAYAN, 1998). The rationale for this practice includes precluding the acquisition of tenurial security by workers and evading the application of labour laws. Thus, workers who are of contractual status are paid minimum wage and are not entitled to the rights and benefits of employees who are of regular status (Diokno, 2000). Given the perceived corporate benevolence of PMC, the company does not veer off from its primary and ultimate objective, i.e., maximising profits. In a case that was filed by several workers against PMC in 1999, the Supreme Court found the company in 'a subterfuge', having imposed contracts on the workers so the company can circumvent the workers' 'right to be secure in their tenure'. The Supreme Court described the company as one 'with vast operations' but is 'not immune from the temptation of circumventing labour laws for the sake of profit' (Supreme Court of the Philippines, 1999).

Weaving work with kith and kin

Philex is, first, an occupational community. Every household head, predominantly the husband, is an employee of the company. The reason for a family residing in Philex is the employment of either the husband or wife (or both). As a policy, non-employees are not entitled to reside in Philex, although this rule has been bent in some circumstances.

The composition of the first group of senior staff in the company had, to a large extent, determined the present configuration of Philex's population. Philex residents represent different ethnolinguistic groups coming from various provinces. The earliest employees were hired from the 1960s through to the 1980s. During this period, individual department managers were in charge of meeting manpower requirements so they themselves were responsible for recruitment. The manager of the construction department recruited workers from Bauang, his own town located in La Union Province. The manager of the mill department recruited workers from Ifugao Province, his origin. Another manager who previously worked for an iron mining company located in Bicol arrived in Philex and soon recruited workers from Bicol Province. This makes Philex, secondly, a community of kith and kin.

The minesite camp is made up of people who arrived in Padcal through their association or relation with someone who arrived earlier. Almost every worker got a job because they knew of someone already working for PMC and residing in the camp. It has been a practice and an expected obligation for a *kabagian* (relative by consanguinity or affinity) or *kailian* (co-villager) to inform a kinsfolk about opportunities for greener pasture. To many relatives living in the rural areas, the added attraction of Philex is its proximity to Baguio, a major educational and commercial centre in northern Philippines. Many unmarried individuals who visited Philex also ended up meeting each other, leading to their marriage.

The 'freebies' in Philex include housing, water and electricity, education for workers' children in the primary school, and highly-subsidised tuition in high school. Among countless families, their privileges that come with being residents in a minesite camp is something they are keen to share with kinsfolk, often as a filial obligation. Thus, bunkhouses have become overcrowded partly due to the presence of extended family members. Husbands and wives have had nieces, nephews, younger cousins, and in-laws staying with them, and most of them study in the company-run schools. Older children who leave Philex to pursue tertiary courses in Baguio are shocked at the life in the city where one pays rent for housing, bills for city services, and fares for public transport. Students who stay in boarding houses in Baguio regularly come 'home' to Philex on weekends. Bringing their laundry to the minesite camp where water and electricity are 'free' is part of the weekend break.

Ranks and privileges

In an occupational community, people do not have the same entitlements because jobs and ranks are different. The differences in privileges, which are most visibly reflected in housing and associated amenities, often serve to inspire some children of miners and other low-ranking employees to aspire for a position higher than what their parents held. A sentiment of a 49-year-old manager expresses the entrenched desire to acquire a status better than his father's:

> I was nine years old when we lived in a bunkhouse in Upper Camp. I often passed by the Main Camp, the area for managers. Every time I walked along the road there, I was awed at the row of big houses that exuded not only power and authority in Philex but a kind of aura that they were somewhat unreachable. The yards of the big houses in the Main Camp were completely fenced and their gates closed. It felt like a lowly miner's son like me does not belong there. As a young boy, I made a promise to myself that I would one day become an engineer and occupy a good position within the company.[12]

A substantial number of employees of PMC are children of former and current employees, and referred to as 'second generation' Philexians. Records at the Human Resources Department show that in 2005, there were 482 employees – representing 18 percent – whose parents had worked or are still working for the company. Among the 70 managers, there were 16 (22.85 percent) who are 'second generation' employees.

As pay and privileges come with one's position in the company, employees of lower ranks talk about one's need for diligence and *pakikisama* (rapport) 'with the bosses' to gain headway in one's position. Nonetheless, it has been a common assertion by older residents that the distinction between top-ranking and lower-ranking employees in Philex pales when compared to the cases of other mining camps. The remarks of a 37-year-old mining engineer who grew up in another minesite camp, highlights an important element of community life in Philex:

> In Philex, the company is the community; the community is the company. As a mine, it has the most modern equipment in mining in the Philippines. As a community, it has good roads, good schools with top-notch teachers, and a medical centre. I grew up in another mining community, but a totally different world. I came to Philex in 1989 to visit my brother. I was amazed at finding out that the way people relate to each other here is very different. The rank-and-file employees mingle with supervisors in the sports grounds, in the canteen, in the streets. This was something that never happened where I grew up. In that mine camp, the staff[13] are often high nosed, tending to look down on downtrodden miners and muckers and their families. Also, the children of the staff ride in a shuttle bus that brings them to school. A label is stuck at the back of the

> bus indicating, 'Exclusively for staff only'. The children of staff do not pay fares. The rank-and-file employees and their children take the jeepney, or walk. Here in Philex, the facilities for recreation and transportation are for everyone. Both the managers and rank-and-file employees are serviced by the bus to and from work.[14]

The blurring of ranks in social relations within Philex is not necessarily seen in positive. For example, a mechanical engineer who grew up in Antamok recalls:

> My father started in 1951 as a miner but after 10 years, he was promoted as a *capataz* (overseer). Later, he was further promoted as a mine haulage supervisor, the position he held until his retirement in 2000 at age 60. We had good housing because of my father's position. After I finished my degree in mechanical engineering, the company hired me because it was its recruitment policy to give priority to dependents of mine employees. It was good living in Antamok because I was given the space and entitlement appropriate for me, an engineer. Here in Philex, the setup of housing is chaotic – miners and supervisors live near each other, which should not be. In Antamok, it was orderly. The area for junior staff was exclusively for junior staff, no lower-ranking persons allowed.[15]

The residents in Philex who have lived for more than 20 years recall that the division between higher-ranking and lower-ranking employees has in fact eroded over the years. All managers are still concentrated within the Main Camp although the vicinity is not off-limits to non-managers. Many supervisors who started as lower-ranking workers maintained their residence in the bunkhouses even after they were promoted.

Administering community affairs

Part of Philex's evolution include the establishment and growth of several organisations initiated by residents. These include, among others, the long-operating Philex Community Credit Cooperative, established in 1965, through which workers and their family members have engaged in systematic and continuous savings program to provide loans to each other at reasonable interest rates.

As a minesite camp, Philex is divided into several areas with each comprising several *purok* (zones). In 1977, the residents organised the Philex Purok Organisation (PPO) composed of officers from each zone. The PPO formally joined Barangay Camp Three Council and PMC to provide more systematic management of the minesite camp and address problems associated with a growing multi-ethnic population. The company held in Upper Camp the first meetings on neighbourhood cleanliness and peace and order. The most pressing challenge in Upper Camp, as the PPO coordinator emphasised, was 'the tribal conflict and factionalism because of provincial differences' where conflicts in the villages tend to be transposed to Philex. Nonetheless, PMC held seminars on topics such as leadership and conflict resolution to curb fighting between neighbours.

Over many years, the PPO has organised regular forums for Philex residents on fostering cooperation for sanitation and cleanliness, prevention of crime, preservation of public safety, and building friendly relations through various social activities such as sports, Bible studies, etc. The company has provided materials and equipment needed in community projects such as camp cleanliness and beautification, minor repairs of cooking and washing sheds, de-clogging sewers, and settling domestic problems among households and neighbours. In every purok, there are *barangay tanod* (peace and order brigade) who had undergone skills training on peacekeeping and security. They coordinate with the company security force in keeping the camp peaceful and orderly, particularly during special occasions[16] that draw large crowds. The PPO has been the company's primary channel of communicating relevant information such as salary increases and special bonuses, and reminders about everyone's cooperation to meet mine production targets. It is not unusual to hear housewives talk about 'ore tonnages', 'Level 782', 'cost reduction', and 'LHD' (load-haul-dump). Since the 1990s, the company, the PPO, and the Barangay Camp Three Council have jointly enforced curfew in Philex.

Philex residents have also organised themselves according to their province of origin. The province-based organisations were set up primarily for socialisation purposes such as rendition of ethnic dances and songs during special events like Christmas, Independence Day, and Easter. These organisations become venues by which residents identify themselves as belonging to an ethnolinguistic group. These same organisations also incorporate in their objectives and activities company-related concerns. An example was the United Highland Association (UHA), the organisation of residents from 'the different highland tribes of Bontoc, Ifugao, Kalinga, Apayao and Benguet working in Philex'. While UHA is no longer active at present due to the retrenchment of some employees and departure of some members, one of its objectives was 'to attain the highest level of employee efficiency and performance through persistent observance of safety and health consciousness'. The organisation's by-laws included a specific provision to encourage members to become exemplary workers of the company:

> [A]ny UHA member who can attain or render three consecutive months without incurring any sick leave, vacation leave, or rest day

> will be given 25 kilos of rice. The award shall be borne by the association through its collectible funds. Meanwhile, any member found stealing company's or somebody's property will be penalised with an amount twice the value of the property. And any member found to have committed public scandal within the company premises will be meted a fine of 50 pesos. All the general fines shall go to the association's coffers to be used for other worthwhile projects that include the fabrication of native attires to be used as costumes for special presentations, purchase of athletic uniforms of the UHA players, provision of sound system for social occasions and the provision of a cooking vat for cañao purposes.
>
> PMC (1980a)

Philex residents coming from Pangasinan Province have sub-organisations based on the respective municipalities where they come from. Their aims include constructing waiting sheds and other public infrastructure in their respective hometowns. Meanwhile, the Ilocano Association, comprising workers coming mainly from La Union and the Ilocos, aim to encourage their members in practicing 'industry, thrift, fortitude, a cheerful resignation to drudgery and an aptitude to survive' (PMC, 1981a). Organisations of residents coming from other parts of Luzon as well as the Visayas and Mindanao had also been active for years.

Given the relative ease in coordinating activities within Philex with much of the structure and process provided by already existing organisations, Barangay Camp Three Council – the political unit whose jurisdiction covers most of Philex – has been able to implement with remarkable success the *Katarungang Pambarangay*. It is an extra-judicial mechanism of settling amicably interpersonal disputes beyond the halls of the formal courts of justice. To implement a just and prompt administration of justice, Barangay Camp Three organised the '*Lupong Tagapamayapa*', the Council of Mediators, involving 20 individuals chosen from community leaders and elders. For the period 1990–1994, a total of 359 cases were filed at the Barangay, with 298 cases – or 83 percent – settled, 34 cases endorsed by the courts, and 16 cases dismissed. This record has placed the officials of Barangay Camp Three in a position of getting invited to other provinces of the Philippines to share their experience. Most of all, Barangay Camp Three was awarded in 1995 by the *Friends of the United Nations* as one of the 50 model communities in the world. The Award was part of a program in honour of the 50th Anniversary of the United Nations. The *kapitan* (local council executive) of Barangay Camp Three, who travelled to New York, USA to receive the award, acknowledged that Philex's being a unique community under the auspices of a mining company, where cooperation and tolerance among diverse peoples are well-entrenched, accounted for Barangay Camp Three's success in setting an exemplary performance in dispute settlement.[17]

The conditions of employment, housing, and service-provision by PMC might appear to be excessively paternalistic and hierarchical from some perspectives, but interviews revealed high levels of satisfaction among employees at every level. While relative economic advantage underpins such attitudes, the fact that these are constructed in terms of widely-held Filipino values about the family, patronage, and regional loyalties ensures that company intrusions into everyday existence are perceived as benefits rather than constraints.

Departures, deferments, and dispersals

In 1992, PMC incurred substantial financial losses. The early 1990s had been a period of significant slump for the mining industry due to the decline of metal prices. Thus, PMC adopted cost reduction measures. These included manpower reduction by offering early voluntary retirement and retrenchment programs (Supreme Court of the Philippines, 2005). In 1993, over 300 individuals left the workforce. Another reduction took place in 1996 'due to consistent losses, progressive debts and ongoing mechanisation program of the company', that saw 519 employees availing of the company's offer of a tax-free early retirement program (PMC, 1996). In 2003, a further reduction entailed the departure of 338 employees (PMC, 2004, p. 4). The company has continued relying on contractual labourers.

With the series of workforce reductions, most of the province-based organisations became inactive because some members left Philex. Many residents talk about the more exciting days of Philex as a thing of the past. A more significant change associated with retrenchment pertains to the anxieties among affected families. While retrenched and retired employees were happy to receive their severance payments, the sentiment among a larger number of families has been one of reluctance if not aversion. In effect, the company has found itself having to deal with social, administrative, and legal challenges.

While huge numbers of retrenched and retired employees gradually departed, scores of others have continued to stay in Philex. The company granted the requests of some families whose children were still studying in the company-subsidised schools to stay until the end of the school year corresponding to the year the workers' employment ended. Other families found ways of legitimising their stay by engaging in retail businesses within the camp. The perceived leniency which the company applied on some families, who were retrenched in 1993 but had not left, was taken as justification by other employees who were laid off during the subsequent retrenchments (i.e., 1996, 2002 and 2003) to also continue residing within Philex. This meant they have also continued occupying their quarters and using water and electricity. However, other defiant employees, like a group of supervisors, challenged the legality of the company's decision to dismiss them from service (Supreme Court of the Philippines, 2005), asserting that they will not leave the camp until a final decision on their case is issued. Incidentally, their court case that was filed in 1993, which was appealed against earlier rulings issued in 2000 and 2001, lasted until August 2005 when the Supreme Court finally issued a decision directing the company to reinstate the dismissed supervisors. For the entire period of 1993–2005, the dismissed workers, who felt vindicated later, stayed in the camp and enjoyed free housing, water, and electricity.

In early 2005, the company had notified non-employee residents in Philex to leave their living quarters or shacks lest their water and power services get cut off and the contents of their dwelling units forcibly brought out. The company carried out an activity which was called 'camp decongestion program'. It involved coordinating with the police from the municipalities of Tuba and Itogon. The program aimed to expel former employees from their bunkhouses and shacks inside Philex. The activity took place in mid-February, and resumed in April. The experience of some company personnel who were part of the decongestion team, and the insights of observer-residents, highlight important facets of a community that has been intimately attached to the company. The Housing Supervisor narrates:

> My job is difficult because it involves direct interaction with people affected by the company's workforce reduction program. They are anxious about their future so they are very emotional and some are extremely angry. They may see us as enemies. That is the reason why we have the police with us. In our decongestion activity, we try to be cautious in our move because we are aware we can be charged with violation of human rights. After we issued the notice of eviction to concerned employees, a few of them heeded to the order that they vacate their quarters, but many others chose to ignore the order. Some people deliberately refused to entertain us. They were not around when we arrived at their quarters. But we had to forcibly enter the rooms, pack up all the items from the smallest to the biggest. It is the inventory of items that eat up our time. We had to carefully list down all items that we put into the boxes, lest we be charged with looting. And then we had to label each box and take the boxes out of the rooms. And finally, we cut the power and water lines. There was one apartment whose lady occupant was awfully furious. It was not easy hearing her endless yelling and curses directed at us. I just answered her, 'Madam, I do not think

> your curses will work because God is merciful to us.' We did not pack up her belongings because we could not stand her yelling. We decided to cut off only her water line, and left her power line still connected.[18]

Both camp residents and outsiders regard PMC as deficient in marking the boundaries of what is permissible or prohibited. A nurse at the Philex's Santo Nino's hospital observes:

> Life in Philex has been one that people have enjoyed. It is not at all easy to leave the place, because it means a change in the lifestyle one has always known. People in Philex have been pampered with many things that are free. People did not have to pay rents for their rooms in the bunkhouses, and bills for water and electricity. The company is at fault. It has not been clear with its rule that former employees should not be residing in the camp. The company has been poor at managing people's expectations. Even before the first retrenchment, there have been many retirees who continued to stay in the camp. Some of them do not occupy company-owned quarters but they built their own shacks within the camp. The time came for the company to decongest Philex because it just cannot continue supporting everyone, but it is now too difficult.[19]

The eviction process that PMC carried out drew controversies in local weeklies in early 2005. PMC was on the news after one retrenched employee led other prospective evictees in seeking action against the company, drawing the attention of the provincial government of Benguet. The provincial board reprimanded PMC's non-conformance to the temporary restraining order issued by the board several months before the decongestion program was carried out. The restraining order aimed to stop the company from evicting the retrenched employees. One board member's remarks (Belinan, 2005) highlight an impression that forms part of public expectations from mining companies:

> How could we take a pro-mining stance when mining companies are doing these things? Philex has already sapped the productive years of its employees and now it wants to get rid of them. It is ironic that Philex was given the child friendly company and the environmental friendly company awards last year.

Residents have first heard in early 2004 that the Padcal mine will close in 2011 if no new ore discoveries are made. The reaction of most residents to this news was one of anxiety. A few older residents took the news as untrue. By May 2008, residents received the management's announcement that the

Padcal mine will carry on operating until 2014, three years longer than previously projected. In effect, some residents felt they may not have reasons to worry because the Padcal mine might not close – a clear negation or defiance against the probability of leaving Philex. The company later revised the closure date to 2017, and then 2020. In August 2011, PMC announced again that the closure date of the Padcal mine is 2022. Some residents have regarded the new date as important in their personal plans. For some, the new date is the ultimate warning that departure from Philex is at hand.

PMC's Final Mine Rehabilitation and Decommissioning Plan (FMRDP) (PMC, 2009) has undergone a few revisions from its original version that was prepared in 2006. The Community Relations Division (CRD) staff had begun conducting a social audit as early as mid-1999 which focused on gathering information about people's awareness of the Padcal mine's closure, expressions of aspirations, expectations, suggestions, and apprehensions about the future. The CRD staff carried out a series of consultations and workshops, all towards preparing a mine transition plan, which formed part of the FMRDP. A much clearer vision has emerged that Philex will become an industrial and commercial centre in the future, and the surrounding sitios with in Tuba and Itogon will serve as satellites for associated ecotourism ventures.

Notes

1 Author's interview, April 2005, Padcal, Tuba.
2 The second company is Lepanto Consolidated Mining Company in Mankayan, also in Benguet Province.
3 Approximately equivalent to US$2.19 million at $1 = PhP42.44 in 2013.
4 Approximately equivalent to US$24.27 million.
5 The studies were carried out by Manila-based CRL Environmental Corporation and *Sentro sa Pagsusuri, Pagsasanay at Pangangasiwang Pang-Agham at Teknolohiya* Corp. (SentroTek).
6 Mr. Henry A. Brimo passed away in March 2007.
7 A *cañao* is traditional festival or ceremony usually to celebrate productive economic activities. The ancestors' spirits are invoked, and thanks are offered to the God *Kabunian*. Animals are butchered and attendees partake of food, and receive their separate shares of meat.
8 *Apo* is a general term denoting respect. It is usually a title for a senior or elderly person, or an honorific term to address the same.
9 Author's interview, September 2004, Padcal, Tuba.
10 During 1984–2000, PMC employees received mid-year bonuses and 14th-month pay bonuses. They received *extra-gratia* in 1984–1988, and 2000–2002.
11 Author's interview, May 2005, Padcal, Tuba.
12 Author's interview, April 2005, Padcal, Tuba.
13 *Staff* is a generic term referring to 'higher-ranking' employees such as managers and supervisors.
14 Author's interview, November 2004, Padcal, Tuba.
15 Author's interview, September 2004, Padcal, Tuba.
16 These include the Philex *fiesta* every January and the Barangay Camp Three *fiesta* every February.

17 Author's interview with Punong Barangay Ignacio Rivera, September 2004, Padcal, Tuba.
18 Author's interview, March 2005, Padcal, Tuba.
19 Author's Interview, Feb 2005, Padcal, Tuba.

References

BAYAN (Bagong Alyansang Makabayan). (1998, 2 May). Newsclippings on the Labor Day (May 1) People's Actions – Workers Hit Contractualization of Labor. Retrieved from www.hartford-hwp.com/archives/54a/193.html

Belinan, J. B. (2005, 27 February). Board members set to grill Philex reps on eviction of former workers. *Zigzag Weekly*, 3.

Diokno, M.S.I. (2000). Human Rights Concerns Arising from Corporate Activities in the Philippines. Retrieved from www.ichrp.org/files/papers/18/107_-_The_Philippines_-_Business_and_Human_Rights_Diokno__Maria_Socorro__2000.pdf

Environmental Justice Atlas. (2015, 24 April 2015). Philex's Padcal Mine, the Biggest Mining Disaster of the Philippines. Retrieved from https://ejatlas.org/conflict/philex-padcal-mining-disaster-benguet-philippines

Gozun, E. G. (2003). Welcome Address – Opening of the National Mining Conference (Holiday Inn Galleria Manila, Ortigas Center, Pasig City). Retrieved from www.mgb.gov.ph/pmwc_nmc_welcomegozun.htm

Kalikasan-Peoples' Network for the Environment. (2013, 28 February). Activists Throw Mud at DENR, Accused Environmental Officials as Lapdogs of Philex. Retrieved from www.kalikasan.net/press-release/2013/02/28/activists-throw-mud-denr-accused-environmental-officials-lapdogs-philex

Lopez, S. P. (1992). *Isles of Gold: A History of Mining in the Philippines*. Singapore: Oxford University Press.

Philippine Daily Inquirer. (2016, 11 December). Philex @ 60: Building Communities, Families. Retrieved from http://business.inquirer.net/221139/philex-60-building-communities-families.

PMC (Philex Mining Corporation). (1958). *Philex Mining Corporation Annual Report 1957*. Pasig, Philippines: Philex Mining Corporation.

PMC (Philex Mining Corporation). (1963). *Philex Mining Corporation Annual Report 1962*. Pasig, Philippines: Philex Mining Corporation.

PMC (Philex Mining Corporation). (1965). *Philex Mining Corporation Annual Report 1964*. Pasig, Philippines: Philex Mining Corporation.

PMC (Philex Mining Corporation). (1980a, January–March). CSR inducts UHA officers. *Philex Cu-Au Bulletin*, *3*(1). Pasig, Philippines: Philex Mining Corporation.

PMC (Philex Mining Corporation). (1980b, January–March). Editorial: Our little world. *Philex Cu-Au Bulletin*, *3*(1). Pasig, Philippines: Philex Mining Corporation.

PMC (Philex Mining Corporation). (1981a, January–March). Philex Ilocano association. *Philex Cu-Au Bulletin*, *9*(3), July–September 1981. Pasig, Philippines: Philex Mining Corporation.

PMC (Philex Mining Corporation). (1981b). *Philex Mining Corporation Annual Report 1980*.

PMC (Philex Mining Corporation). (1995). *Philex Mining Corporation Annual Report 1994.*

PMC (Philex Mining Corporation). (1996, January–February). Company offers early retirement. *Philex Cu-Au Bulletin, 24*(1).

PMC (Philex Mining Corporation). (2004). *Philex Mining Corporation Annual Report 2003.*

PMC (Philex Mining Corporation). (2006). *Message to Stockholders (In 2005 Annual Report).*

PMC (Philex Mining Corporation). (2007, 16 April). Philex Mining Corporation 2006 Annual Report (SEC Form 17-A for Period Ended December 31, 2006). Retrieved from www.pse.org.ph/html/ListedCompanies/pdf/2007/Px_17A_Dec2006.pdf

PMC (Philex Mining Corporation) (2008). Philex Mining Corporation Annual Report 2007. Retrieved from http://globaldocuments.morningstar.com/documentlibrary/document/27b4efb80e7701e8.msdoc/ original

PMC (Philex Mining Corporation). (2009). Final Mine Rehabilitation and Decommissioning Plan. Retrieved from http://contracts.ph-eiti.org/supporting-documents/Philex%20Mining%20Corporation%20-%20MPSA%20No.%20276-2009-CAR,%202009/FMRDP_Philex%20Mining.pdf

PMC (Philex Mining Corporation). (2010). Philex Mining Corporation Annual Report 2009: Mining for the Future: Retrieved from http://www.philexmining.com.ph/2009/06/15/annual-report-2009/

PMC (Philex Mining Corporation). (2012a). Corporate Profile. Retrieved from www.philexmining.com.ph/about-us/corporate-profile

PMC (Philex Mining Corporation). (2012b). Philex Mining Corporation Annual Report 2011: There Is Life in Mining. Retrieved from www.philexmining.com.ph/userfiles/file/Philex%20Mining%20Annual%20Report.pdf

PMC (Philex Mining Corporation). (2013a). Philex Debates Critics on Padcal Issue. Retrieved from www.philexmining.com.ph/newsroom/press-release/philex-debates-critics-on-padcal-issue

PMC (Philex Mining Corporation). (2013b, 27 February). Value Nation: Philex Mining Corporation Annual Report 2012. Retrieved from www.philexmining.com.ph/userfiles/file/PMC%20Annual%20Report%202012.pdf

PMC (Philex Mining Corporation). (2015). Philex Mining Corporation Annual Report Part 1. Retrieved from www.philexmining.com.ph/userfiles/annual_reports_2014/PMC%202014%20Annual%20Report%20Part%201.pdf

PMC (Philex Mining Corporation). (2017). Philex Mining Corporation Annual Report 2016: Nurturing Life Shaping Dreams. Retrieved from http://www.philexmining.com.ph/2017/06/29/annual-report-2016/

Supreme Court of the Philippines. (1999, 10, August). G.R. No. 125132. Philex Mining Corporation vs. National Labor Relations Commission, Rosella Austria, Lina Tamondong, Cornelio Borja, Jr. and Gerald Dela Cruz. Retrieved from www.supremecourt.gov.ph/jurisprudence/2005/aug2005/147756.htm

Supreme Court of the Philippines. (2005, 9 August). G.R. 147756. Ariola vs. Philex Mining Corporation. Retrieved from www.supremecourt.gov.ph/jurisprudence/2005/aug2005/147756.htm

Terso, J. O. (2017). *The Test of Fire for All of Us: My Community Development Adventure During the 2012 Tailings Spill.* (unpublished mimeo). Padcal, Tuba.

4 Company relations with communities around the Padcal mine

> Being indigenous is a blessing because there are entitlements, but I am saddened that my *kakailiyan* (co-villagers) are once again disunited – fighting against each other because of the royalty payment that the organisation received from the company. We thought that putting up an 'IPO' (indigenous peoples' organisation) will help us economically and socially, but instead, bigger problems came up. The membership of the IPO multiplied when people saw that money was pouring. There are four IPOs now, instead of one. The original IPO was made up of more than 60 members, but now I am not sure how many hundreds – maybe 600-plus. There are four or five different sets of leaders all claiming to be the true representatives. It is the original IPO officers that signed agreement with the company and NCIP in 2009, but everything became complicated. The NCIP ordered the bank not to release funds to IPO representatives. Now, the funds and the good purposes of royalty are frozen.[1]
>
> Dolores Amadayo (pseudonym)
> Daughter of an IPO-APSSOL member

The mine area of Philex Mining Corporation (PMC) is located within a district of indigenous and migrant populations. The company has found itself dealing with people individually, by family or clan, and often as members of an organisation representing indigenous interests and claims to compensation. The people around the Padcal mine have gained awareness and experience in asserting their claims to benefits and entitlements from mining. The relations of PMC with communities around its mine feature the process of claim-making by indigenous families and other residents of outlying sitios at a time that could not have been more opportune but simultaneously a little too late. The actions of leaders and representatives of claimants have gravitated on the impending closure of the Padcal mine following the company's public announcement in 2004. The claims for compensation to damages and disruption of traditional livelihood have, to a large extent, shaped the directions of PMC's social development and management programs (SDMP), as the company aims to demonstrate and publicise its programs in environmental management and social responsibility.

The surrounding sitios

There are more than 20 *sitio*s around the Padcal mine and the minesite community (Philex). A sitio comprises small and sparse nuclei of households. A group of sitios make up a *barangay*, the smallest administrative-political division in the Philippines. The Kalanguya who arrived in the Padcal area in the 1950s have settled in the northwest side of Philex. They trace their origins to Tinoc, a municipality of Ifugao Province. They are upland farmers growing root crops and vegetables for home consumption although a small number of households sell some of their products such as bananas, ginger, beans, corn, and cabbage in Philex. The far eastern side of Padcal is settled by Ibaloy families engaged in small-scale gold mining. They trace their ancestry to Beckel, an agricultural barangay in La Trinidad municipality. Around this area are small clusters of Ibaloy families who were traditional cattle raisers and traders long before PMC arrived in Padcal in the 1950s.

Scattered still in the eastern side of Padcal are newer settlements of small-scale gold mining groups including migrants from other provinces within the Cordillera region such as Ifugao, Kalinga, and Mountain Province, who were attracted to the 'gold rush' in Taloan in the early 1980s. They intermarried with the indigenous Ibaloy and settler Kankanaey, and had since resided in the area. The newer small-scale gold mining groups occupy Nay-en, Sais, and Sinko. Older and established Kankanaey and Ibaloy families who engage in both mining and farming reside farther down in sparse hamlets like Palunupon and Piaki where people from Nueva Vizcaya and a few from Mountain Province have settled since the 1960s. Meanwhile, along the Baguio-Philex Road towards the east side of Padcal are Kankanaey families who established residence as early as before the Second World War. Many of them came from Buguias in the northern part of Benguet. While they cultivate vegetables in their backyard farms, their main livelihood is small-scale gold mining.

A population survey which PMC carried out in 2002 indicates that the surrounding sitios consist of 945 households comprising a total of 5,414 individuals. These settlements fall under two local councils, namely Barangay Camp Three within the municipality of Tuba and Barangay Ampucao within the municipality of Itogon. Having two administrative divisions that govern the sitios often brings up equity issues concerning short-term and long-term social investment programs that PMC provides to surrounding communities.

Claiming benefits and entitlements from mining

In 1999, some members of an Ibaloy clan, with Kalanguya and Kankanaey families, established an organisation that they named TIHCAPOSI which stands for 'The Indigenous Host Communities and Philex Outlying Sitios Incorporated'. Organising TIHCAPOSI was initiated by a schoolteacher whose grandparents were the owners of the land acquired by PMC for the construction of the mill in the early 1970s. The schoolteacher who also served as TIHCAPOSI's first president explained that she had read the Mining Act of 1995 in relation to her graduate thesis. She described what she understood from the law: indigenous peoples in host and surrounding communities of mining operations should benefit from the proceeds of mineral development. Available documents of TIHCAPOSI state the organisation's objective: 'to ensure that the indigenous peoples inhabiting the Padcal area and vicinity are granted the rights they are entitled to'. TIHCAPOSI aims 'to be a channel for advancing the indigenous peoples' aspirations' (TIHCAPOSI, 1999a, 1999b).

Within the sitios, people generally acknowledge that PMC has brought them the social and economic benefits of being connected to the outside world. This was realised through the Baguio-Philex Road that PMC built in the 1950s which involved clearing, widening, and sealing an old pre-Second World War logging trail. The road has provided the sitios access to and from Baguio. Until 2013, this has been the only reasonably passable route in the southeastern part of Tuba. Through its Social Development Management Program (SDMP), PMC improved a separate dirt road passing along the sitios of Alang, Torre, and Balayan that exits to Kennon Road, a national highway. While residents in the surrounding sitios recognise the developments that PMC has brought to the area, many of them consider the provisions they receive from PMC as a pittance compared to the privileges enjoyed by company employees and their families. There are feelings of resentment within the sitios surrounding the minesite camp about the large number of migrants who make up PMC's workforce. The local Ibaloy are quick to distinguish themselves as *Taga-Ditoy* ('of this place') from the *Taga-Baba* ('from the lowlands' or from provinces outside the Cordillera) or *Taga-Sabali* ('from other places'). For instance, in my conversation with an Ibaloy small-scale gold miner, his remarks express some aversion towards non-locals:

> If you count the number of employees of the company during the past many years until the present, there is hardly any Ibaloy. Out of the thousands of employees that the company hired over many years, only 27 were *Taga-Ditoy*, and they had not stayed on. Nepotism in the hiring of workers has been the practice within the company. For all the years

> that the company has existed, it is the lowlanders and the *Taga-Bontoc*[2] who have had the jobs. We locals could not get into any job because the priority for hiring are always the employee dependents. The practice of *palakasan*'[3] and '*kumpare* system'[4] is rife.[5]

PMC's Human Resource Department (HRD) personnel explain that a few Ibaloy and Kalanguya had been employed in the mine, but they resigned only after a few years. According to HRD, the lowlanders and other employees have demonstrated commitment to work for the mine and many of them had done so until retirement. When the company started operations in the 1950s and began recruiting workers, the Ibaloy in Padcal and surrounding areas occupied themselves with farming and cattle-raising and were not available for jobs at PMC. Similarly, the Kalanguya were busy growing upland crops in their swidden farms although a small number of Kalanguya men worked shortly as drilling aides during the earlier part of PMC operations in the late 1950s towards the 1960s. By the late 1970s, farming and cattle-raising had become less significant in Tuba and Itogon as land for pasture became more restricted. At best, the jobs that were made available for the local population are contractual and casual work. One Kankanaey research participant who once worked for the company noted, 'there is no room for us; the permanent jobs are all filled up; the people working for the company have no intention to leave their posts until they either die or retire.'[6]

Claiming compensation for damaged land and lost water

Since the late 1990s, PMC's legal department together with the Community Relations Division (CRD) has been dealing with individual families who have land claims within and around the company's area of operation. PMC deals with two groups of council officials, and across two tiers of local government – i.e., barangay and municipality levels – in permitting concerns as well as identifying, planning, and implementing SDMP. Considering that the local government is required to monitor the environmental programs of large-scale mining companies, the Multi-partite Monitoring Team (MMT) and the Mine Rehabilitation Fund Committee (MRFC) for the Padcal mine include participants representing Barangay Ampucao and Barangay Camp Three.

Among the more than 20 sitios the Padcal mine, Alang is one where residents have long-drawn-out complaints against damaging effects of PMC's operation. Alang used to be a settlement of over 30 predominantly Kalanguya households whose primary livelihood is upland farming. Incidentally, the Santo Tomas II ore body of the Padcal mine lies underneath the farms in Alang. In the early 1970s, the primary school in Alang had to be relocated from its original site to Torre, an adjacent sitio, due to the drying up of springs which served as sources of water for households and the school. Kalanguya families attribute the water depletion to PMC's decades-long operation. A Kalanguya elder explains that the open pit excavation of the company in 1958–1960 was not the only cause of the loss of water in Alang, but also the continuous block-cave mining of the company since early 1960s until the present. The observed reduction of water from springs not only affected the school, but more critically, the irrigation requirements of farms. In 1998, PMC paid compensation to 16 families in Alang whose crops, farms, and house lots that its mining activities affected. Each of the families claiming for compensation entered into individual memorandum of agreement with the company. Some elders describe how a few individuals acted as representatives and negotiators in behalf of some families in the settlement of claims, then taking a share from the payments. Several former residents in Alang regret that relations between relatives and neighbours had been severed due to the shrewdness of a few negotiators, Kalanguyas themselves, who were perceived to be imposing their own wishes not those of the families affected. The compensation payments made some families millionaires overnight while others were unhappy because PMC found their claims as non-legitimate.

From the 1980s to late 1990s, the families residing in Alang had gradually left. Some moved to bordering sitios while others acquired accommodation in Philex. Many others joined relatives in Nueva Vizcaya, a province adjacent to Benguet, where they invested their compensation money on vast rice lands. A few others went to pursue commercial vegetable farming in Loo, Buguias, an agricultural town. The departure of people meant the surrender of the last hope of reviving Alang. It marked the abandonment and eventual dissolution of what was once a flourishing agricultural community. The recollection of Mina, a woman in her late 40s, expresses profound sentiments about Alang:

> Alang, as you have visited, is at a high elevation. Such location gives the place a climate that is perfect for temperate vegetables. Potatoes, cabbages, beans, and many more grow very well in our farms as long as we had water for irrigation. But due to the drying up of springs, we are not able to continue farming. We had to look for other means to survive. People leave the place not because they wanted to. Those whose properties were affected were indeed paid their compensation, but seeing them leave Alang breaks my heart because this meant the disappearance of the homeland where we were born and grew up. Alang is simply a memory now.[7]

At the far southeastern end of Padcal is the location of company tailings ponds. One began construction in 1973 and was completed in 1981. The second was built in 1983 and began operating in 1992. Constructing these tailings ponds caused the submersion of pasture lands. Various families have revived their claims for compensation for these lands when they heard about PMC's closing. Some of these families are original Ibaloy settlers in the area while others are migrants who arrived during the 1950s through 1970s. Family members having land claims in the submerged areas, referred to as Dam Two and Dam Three, explain that they received payments from PMC for the lost crops, physical structures, and investments made on the land, but not the value of the land itself. On the other hand, interviews with the company's legal department suggest that all settlements were completed with the affected families. The issues that remain unresolved which have been lingering since the 1990s pertain to payment for the lands. Incidentally, the tailings pond areas fall under forestland and therefore non-alienable and non-disposable for other land use. Being forestland, the owner is the state or the government of the Philippines and not families. Compounding the controversies are conflicts among more than 15 families whose claims on an area of over 30 square kilometres overlap each other. The persistent claims for land payments, despite lands classified as part forestland or watershed, evolved because of the unyielding claimants and feisty endorsement of some politicians. To 'buy peace' with the claimants, PMC willingly settled all payments for the land, provided there is no overlapping claims among the families. The views of a Barangay Council official point to ambiguous land classification vis-à-vis actual land use as complicating factors in compensation matters:

> The problem in Dam Three area is that many families are demanding the company to pay for their land. However, the company insists that all settlement had already been done. Rightly so, the affected crops and houses had been paid. The families are relentless in insisting they get paid compensation for the land, on top of separate payments for their crops and other things. The company eventually paid all claims, but the land is forestland. Anyone can verify this at municipal assessor's office. Of course, anyone can also correctly argue against the company that in the first place, the tailings pond should have never been built on forestland.[8]

PMC officers explain that the location of the tailings ponds was excluded from the Central Cordillera Forest Reserve under Proclamation 304[9] signed by former President Ramon Magsaysay in June 1956. Meanwhile, a separate group of claimants in the area allege that their crops had not been paid for. However, the company consider these individuals as illegitimate claimants. The remark of a PMC officer reflects his censure of what he considers chicanery employed by some individuals to gain compensation:

> When we checked the vicinity of Dam Three, we found that there were hundreds of newly planted coffee trees and sweet potatoes. They were easily detectable to have been freshly planted. I am not ignorant in farming because I did farming when I was young. That is why I know how a real farm ground looks like. In the case of those crops which we found at the Dam Three area, one can easily pull the coffee trees and sweet potatoes; they were not genuine. The funny thing is that these same crops were among those that were already counted and paid for when the company carried out an assessment for compensation payments in Alang.[10]

Meanwhile, the sentiment of one claimant reveals another dimension of claim-making. It explains why PMC's legal department and community relations personnel have painstakingly put together specific criteria for examining the validity of claims:

> Life is difficult and one must find ways to earn income. Many times, finding money is a matter of winning and losing. If you were me, you would better try the trick because . . . *no ngay nagbalin* ('what if it worked')? Considering that the company has money, it is not wrong if the company will give part of its wealth to poor people like us. If your trick did not work, it's fine. But it is your own loss if you did not do anything at all. At least, we tried.[11]

Santa Fe is another sitio where residents ascribe the depletion of their water to the company's drilling activities in previous decades. The residents, whose main livelihood activities include farming and gold panning, describe how their water sources had been drying up. They later joined in establishing an organisation called ASKD 123 which represents families who reside and have claims in Albian, Santa Fe, Keddet, and the areas of Dam One, Dam Two, and Dam Three. They also joined the group IPO-APSSOL which is made up of residents in Alang, Pokis, Sabiang, Sta. Fe, Oliba, and Loacan.

Agreements and mechanisms in settling claims

Part of the company's scheme of settling claims advanced by families whose properties were adversely affected by mine operations include specific privileges. For example, priority is given to members of affected families in hiring contractual or casual jobs. PMC however put forth at the outset the caveat that the applicants should be technically qualified for the job. Nonetheless, this rule has been bent, and some claimants have been hired to fill available jobs. Access to livelihood such as gold panning in company-designated areas, and subsidised secondary education for children, are additional forms of compensation. The claiming families have priority in PMC's awarding of minor contract projects. The families who are permitted to reside within the minesite camp have been granted free electricity and water services.

To establish criteria for assessing whether claims are legitimate or not, PMC has fastidiously delineated what it considers as the physical ambit of influence of its operations. In physical terms, the company plotted this area to be 'more or less five-kilometre radius from the mine, extending 10 kilometres to the south'. PMC's CRD developed this rule together with other criteria and delineated the impact coverage of PMC's operations to include low-lying flat lands where several families reside, and hill slopes where some farmers undertake their agricultural activities. Other areas include the location of structural and mechanical installations for the mine, mill, and tailings storage facilities. The CRD staff defined 'areas of influence' and identified families in specific areas with whom to negotiate regarding compensation.

With many people from the sitios claiming compensation for loss of crops, houses, relevant structures, and land, the company had to categorise its areas of influence into different impact zones: a) high impact, b) medium-to-high impact, c) low-to-medium impact, and d) low impact. The bases of classifying the zones include:

a) area proximity to the mine operations, facilities, and utilities
b) effect of operations on demography, employment, market of products, utilities, social services, and delivery systems
c) accessibility, which includes road networks and/or transportation facilities
d) actual effect of mine closure on the bio-physical environment and ecosystem

(PMC, 2003)

As a rule, PMC has maintained two levels of engagement with the sitios around Padcal mine and Philex. First, the company made agreements mainly with the affected families within the *high impact* and *medium-to-high impact* areas. Hence, a few families got the prerogative of being apportioned a sluicing area where they can pan for gold from mine tailings. Another group of families was granted priority to manage project contracts for the company. These arrangements involve formal legal documents such as Memoranda of Agreements (MOAs) stipulating the obligations of both the company and the families to each other.

The following cases help illustrate the first level of engagement:

Adela Padanos, contractor for civil works

I married Nestor, one of the sons of Dencio, the patriarch of a key family who controls a huge hectarage of land within Dam Two and Dam Three. Since our marriage in 1984, our main economic activity had been gold panning. A garden patch on the hill served as our supplementary source of vegetables and fruits. Our homestead and farms bore damages when Philex constructed its second tailings pond which we call Dam Two. As a form of compensation, the company provided us a canal – along which mine tailings from Dam Two would flow. We sluice the tailings in the canal, from which we derive gold. It is a big help. In 1999, residents within the Dam Two vicinity abandoned their houses due to the rising level of the tailings. The company paid for our property. The settlement of damages involved engaging us for contract projects of the company, an opportunity I seized. Since April 1999, I have been continuously undertaking small civil works for the company, which include construction and repairs of buildings and stone walling of ground slopes. The costs of the projects that I have managed range from [PhP]30,000 to [PhP]40,000.[12]

Auria Dantag, contractor for civil works

I was born in Alang but my family had to move to Philex because we could no longer continue farming. As part of the compensation package, the company provides electricity and water, free of charge, in the house where we are now staying. In 1998, the manager of CRD encouraged me to apply for certification as a contractor so I can avail of company project contracts. I have had quite a few projects since then although I wish the *kapitan* should not be grabbing project contracts, especially the bigger ones, to allow us smaller contractors to earn more. In 2004, I had three projects. One was worth PhP35,000 for the repair of bunkhouse awnings. I accomplished the job within eight days. The total cost for the project was PhP21,000 which covered wages for four labourers. What I was left with PhP14,000, which was a great profit. Another project which costed PhP21,000 involved the concreting of a pavement. I had to hire a concrete mixer for PhP7,500 and three labourers. I was left with PhP4,000 which was still good. I also want to sell scrap iron, a new venture in which many individuals in Philex are engaged. I came to the guest house to look for Sandra, a distant relative of the wife of the

> VPO (Vice President for Operations). I asked her to accompany me to see VPO. I need his signature to make me a certified scrap iron buyer. I am glad that the VPO is Igorot like us. If not, I will never have the smallest courage to go see him.[13]

The small projects being contracted out by the company have allowed a few households to diversify their livelihood. It is also notable that most of the project contractors are women. Men are not attracted to take small contract projects which they regard as 'not worth the trouble'.

The second level of engagement is with communities within all impact zones. The company undertakes the projects under this engagement through the SDMP, a comprehensive five-year plan of a mining company to implement community programs and projects based on consultations with the host and neighbouring communities, as required by the Mining Act of 1995. Ideally, the SDMP should be utilised 'to provide opportunities for a self-sustained livelihood thus decreasing dependency on the benefits derived from mining companies' (MGB, 2005).

The SDMP initiatives of PMC focus on four issue-areas labelled as HELP, an acronym that stands for health, education, livelihood, and physical infrastructure. The projects are directed not only towards enterprise development but also improving the delivery of basic services. Health projects include establishing cooperative drugstores in remote sitios and providing training for housewives on primary health care, nutrition, and sanitation. In addition, residents in the surrounding communities avail themselves of free medical and dental services at the company hospital. The company also conducts medical and dental missions, with participation by staff from the company's Santo Nino Hospital, and involves distribution of medicines for primary health care to families. During these activities, a specialist conducts circumcision to young boys who wish to avail of the service.

Regarding education projects, the TIHCAPOSI has managed to negotiate with PMC to grant scholarships to high school students coming from the surrounding sitios. Thus, in 2001–2005, 44 high school students received subsidised tuition similar to entitlements accorded to children of PMC employees. There were 11 university students receiving full scholarships from the company in 2002–2005. In 2007–2015, there were 250 graduates from university courses, all fully funded by scholarships from PMC. The school year 2016–2017 has 107 university students pursuing various courses under PMC scholarship. Another program which is attributed to the efforts of TIHCAPOSI is the participation of residents in the surrounding settlements in training programs that PMC used to offer only to its employees.

PMC's livelihood and employment enhancement initiatives under the SDMP are wide-ranging. Projects include organising new cooperatives as well as reviving and strengthening old ones, seasonal jobs for agroforestry projects, and skills training in auto-mechanic and welding for youth and unemployed men, broom-making and honey-raising for housewives, and bookkeeping and management skills for officers of community cooperatives. The public infrastructure projects include major construction works such as water systems and road works. Smaller projects consist of sealing pavements, footbridges, concreting foot trails, and repairing water pipelines.

The hospital and schools in Philex serve schoolchildren of non-employees who are from the neighbouring sitios. Their tuition fees in the secondary school were not subsidised and medicines in the hospital were not free, until late 2004 when the SDMP began implementation. I met Melania, a young mother in her mid-20s when she came to the CRD office. Her account presents the idea of claim-making and the need for rules to test the validity of claims for entitlements:

> **Melania, young mother from a neighbouring *sitio***
>
> I came to CRD office to get an important piece of paper regarding my two-week-old baby who is ill. She is having colds and has been having trouble sleeping so I decided to come to the Santo Nino Hospital in Philex. I come from Dalupirip, a far-flung sitio of Itogon but my in-laws are from Bito, a sitio of Barangay Ampucao. It took more than 20 minutes to travel by a jeepney from Bito to Philex but I walked for half an hour from Bito to the main road. At the hospital, a nurse told me that I am not entitled to medical care because neither I nor my husband is an employee of the company. I fully understand that the hospital is for company employees and their families only. However, I also learned from some people that we in the sitios surrounding Philex could now be accommodated for medical care at the company hospital. I told this to the nurse. She explained to me that a requirement she needs is a certification from CRD that I am indeed a genuine resident in one of the surrounding sitios. The people working at CRD know about this requirement. They have a record of the people who are genuine Ibaloy and Kankanaey. They would find out that I am telling the truth.[14]

Preferred social development projects

The local government, through the participation of barangay officials, involves itself in the yearly planning workshops for community development which the CRD facilitate. The idea of local government participation is to help align the SDMP with local development interests. Among the four types of SDMP projects, it is public infrastructure that draws the highest interest of community leaders and local government officials. The relatively greater attention emanates from the fact that infrastructure projects such as roads, waiting sheds, water tanks, and footbridges are physically visible. These types of projects effectively constitute monuments that would reflect the 'benevolence' of the leader who obtained them.

Local government officials have a specific tenure of office and are eager to attribute to themselves projects that are highly conspicuous as these would become useful campaign platforms for subsequent local elections. Infrastructure projects also involve relatively larger finances than other SDMP projects such as skills training or conducting medical missions. Infrastructure projects imply an income source for some individuals because implementation requires some workers. To the workers who themselves must, as a rule, be residents within the surrounding communities, a road construction project means employment for a period that can range from several weeks to a couple of months or longer. To the project contractor who is almost always a local government official, a road project is an effective means to build or maintain patronage with people recruited for the work required.

Indigenousness, locality, and identity

With the approaching closure of the Padcal mine becoming a preoccupying concern, there have been explicit expressions of regret from individuals in many of the sitios about the SDMP not being available in previous decades. They lament that during the period when the company was in its most profitable days, the surrounding communities did not receive the benefits they now can have. Such sentiments are expressed in the context of their perception about what Padcal might look like after the company ceases operating and leaves the place. A remark from a leader of the TIHCAPOSI highlights identity and belonging as important in claim-making:

> The lowlanders and many others in Philex have enjoyed the fruits of the mine. We have not, yet this is our place. Our ancestors did not assert their rights because they did not know what they were entitled to. That's what we are doing this time – asking the company to give us what we should have had in earlier years. It is a bit too late for us to have come together and claim our long-denied privileges because the company is in its twilight years now. Nonetheless, we want to get something somehow rather than nothing. The company will leave Padcal forever, and this place will become forlorn – with us left with nothing. So, we must receive our entitlements now.[15]

Within the occupational community of Philex, there has been an increase in the number of families coming from the surrounding sitios. PMC have granted some families free accommodation with water and power utilities. Other families have been allowed to build their private cottages within the camp. There has been a growing awareness among Philex residents that the people from the surrounding sitios are the rightful claimants of Padcal. The literal term 'claimants' has become an appellation exclusively for the Ibaloy, Kankanaey, and Kalanguya in both Camp Three and Ampucao. The label attaches with it the ideas of entitlement for project contracts and the privileges previously accorded only to PMC employees and their families. Being a 'claimant' – a term which people from surrounding sitios now consciously refer more often to themselves – is also a declaration of a social boundary that has become distinct and relevant as the mine is moving towards closure.

Issues of who is an 'indigenous person' or 'claimant' have been a concern to PMC because in various instances, some Ibaloy, Kankanaey, or Kalanguya bring their nephews, nieces, or other relatives from other towns or provinces expecting priority treatment from PMC for casual or contractual jobs. As a mechanism for qualifying the authenticity of claims, PMC has established an extraordinarily comprehensive record of genealogies of all clans and key families in every neighbouring sitio around Philex. The record of detailed genealogies indicates ancestries to as far back as informants and available documents could point. With the guidance of a retired 88-year-old Belgian priest, Rev. Herman Flameygh, who had lived and served as parish priest in Philex from 1974–1994, the CRD staff have painstakingly put together information on births, baptisms, and marriages pointing to *bona fide* claimants based on consanguinity and affinity, and interviews with families to update births and marriages. The CRD staff have become remarkably well informed about names of key families and how people are related to each other. A key strength of the CRD has been its staff of mainly Philexians and a former resident of Alang who have been well-exposed to the difficult community relations issues PMC faces daily. CRD officers have been seasoned in organising rigorous workshops associated with the mine transition plan and the vision of a post-closure Padcal.

Concerning claims of people about being local and indigenous, the CRD staff have faced the task of testing the legitimacy of claims. They have relied on the voluminous genealogy record which has become a frequently opened and reviewed material. PMC also had to set a cut-off year for determining who are genuine inhabitants in Padcal and its surrounding sitios: 1954, the year before the company arrived in Padcal to undertake exploration. Thus, PMC does not consider as genuine claimants those who may be presently residing in the surrounding sitios but arrived in the area after 1954.

Starting 2013 towards 2014, there have been some changes of staff within the CRD. The effect of being put under a Manila-based manager and having senior officers who are non-locals and non-speakers of the vernacular, compounded by their being non-participants in earlier processes and activities pertinent to SDMP, have become visible. Some residents in sitios within Ampucao observe that the criteria for assessing impact zones and approving priority projects for SDMP are no longer strictly followed. Others say the non-local officers are plainly distant, carrying out SDMP as *tungpal bilin* ('compliance's sake'), not from an appreciation of the situation and the complications involved. One Kalanguya mother observes, 'officers coming from do not know the whole history of the demands of the communities from the company, much less the culture of indigenous peoples in this area.'[16]

Asserting free, prior, and informed consent

PMC's lease contract to mine the Santo Tomas II area expired in January 2007. Thus, the company needed to renew its contract of lease. It was opportune for former residents of Alang to raise once again, and with rather more resolve, their demand for compensation for damaged lands and lost water. The Secretary of the DENR approved the renewal of PMC's lease contract to carry out a mineral production sharing agreement (MPSA) on the condition that the company acquires the free, prior, and informed consent (FPIC) of the communities affected and the endorsement of the local government. The barangay officials of Camp Three and the municipal officials of Tuba gave the needed endorsement. In protest, residents of Alang and neighbouring sitios held in August 2007 a picket against company operations. The picket leaders questioned the DENR's approval of the company to continue operating without consulting the communities affected. To solidify and advance their interests as indigenous peoples, residents of Alang and other sitios established in 2008 the organisation IPO-APSSOL (Indigenous Peoples Organization of Alang, Pokis, Sta. Fe, Sabian, Olibba, and Loacan), a formal group that represents their respective sitios, and is duly registered and recognised by the National Commission on Indigenous Peoples (NCIP).

In a community dialogue convened in August 2010, a lawyer from the NCIP who was present explained that PMC does not need FPIC to renew its contract of lease. He noted that the Indigenous Peoples' Rights Act was passed in 1997 but the Act's constitutionality was resolved only in 2002. The MPSA granted to PMC was issued by the DENR after the company complied with the requirements including proof that the area being applied for mining does not have any patented claims (Baguio Midland Courier, 2010a). The MPSA's approval requires the endorsement of two local government units, the barangay officials of Camp Three and the municipal officials of Tuba, which the company already obtained.

Royalty to indigenous communities

On 14 January 2008, PMC executed a Memorandum of Agreement (MOA) with IPO-APSSOL. The NCIP was also a party to the MOA, as required by law. The MOA stipulates that the indigenous peoples/indigenous cultural communities (IPs/ICCs):

> gave their free prior informed consent to the Company's application for MPSA and to the continued operation and development by the Company of the areas covered by the application, as well as all related, collateral, incident, indirect projects/activities of the Company.
>
> (PMC, 2008, p. 4)

The royalty funds to IPO-APSSOL is 1.25 percent of the company's gross output, as required under the Implementing Rules and Regulations of the Mining Act of 1995.

IPO-APSSOL received its royalty of PhP142 million for community development projects in 2008 and PhP164 million in 2009. The organisation should continue receiving royalty during the remaining lifespan of PMC in the area (Baguio Midland Courier, 2010b; Northern Philippine Times, 2009). In the settlement of all claims for damages, PMC agreed, on the request of IPO-APSSOL officials and local elders, to advance PhP60 million in January of the first year, followed by another PhP60 million in January of the second year. The withdrawn funds were to settle all remaining individual damage claims of the community inside and outside the MPSA area.[17] However, during 2010–2011, there were bitter disputes among IPO-APSSOL members over the royalty, resulting to calls for an audit of how the funds were disbursed, and demands for election of new leaders. Regrettably, what followed were court battles. Throughout these developments, the membership of IPO-APSSOL multiplied as more families claim they too are IPs /ICCs. At the same time, there emerged four sets of officers declaring themselves to be the legitimate representatives of IPO-APSSOL. In effect, NCIP issued a restraining order that barred the bank from releasing funds to IPO-APSSOL. The NCIP wants to deal with one IPO-APSSOL, not four.

Under existing NCIP policy (NCIP AO No. 2012–3, Sections 61 & 62), the ICCs/IPs must formulate a Community Royalty Development Plan for the management and use of the royalties they receive. IPO-APSSOL did not use the funds supposedly for programs and projects that will benefit the people it represents. My field interviews yielded claims and counterclaims of individuals as having received or not received money. Until the present, there appears little hope that the worsening in-fighting within IPO-APSSOL and among their leaders and local elders will be resolved soon. NCIP attempted to validate the composition of council of elders of the four IPOs but this attempt was opposed by the members themselves. Meanwhile, PMC as mandated by law continues depositing into the designated trustee bank the royalties due to IPO-APSSOL.

Notes

1 Author's interview, February 2017, Baguio City.
2 *Taga-Bontoc* refers to 'people from Bontoc' which is generally but imprecisely applied to those coming from the Mountain Province.
3 *Palakasan* refers to the practice that involves selecting a person such as a job applicant based on his or her affinal, consanguinal, or social relations with an influential person.
4 *Kumpare* refers literally to 'fellow godparent'. Associated terms are 'buddy' and 'mate'.
5 Author's interview, February 2008, Padcal, Tuba.
6 Author's interview, March 2005, Market Area, Padcal, Tuba.
7 Author's interview, December 2004, Padcal, Tuba.
8 Author's interview, April 2005, Talnag, Ampucao, Itogon.
9 This law is entitled 'Excluding from Operation of Proclamation No. 217 Dated 16 February 1929, Certain Parcel of the Public Domain Situated in the Barrios of Isheng, Yabua, and Pokes, Municipalities of Tuba and Itogon, Sub-Province of Benguet, Mountain Province, and Declaring the same Open to Disposition Under the Provisions of the Mining Act.'
10 Author's interview, January 2005, Padcal, Tuba.
11 Author's interview, May 2005, Market Area, Padcal, Tuba.
12 Author's interview, January 2005, Ampucao, Itogon.
13 Author's interview, February 2005, Upper Camp, Padcal, Tuba.
14 Author's interview, April 2005, Padcal, Tuba.
15 Author's interview, March 2005, Padcal, Tuba.
16 Author's interview, August 2016, Padcal, Tuba.
17 Author's personal communication with PMC Legal Department. 19 April 2017.

References

Baguio Midland Courier. (2010a, 29 August). IPs, Benguet Dads, LGUs Want Philex to do FPIC. Retrieved from http://baguiomidlandcourier.com.ph/benguet.asp?mode=%20archives/2010/august/8-29-2010/beng3.txt

Baguio Midland Courier. (2010b, 31 October). Mine Firm Gives Assistance to Tuba Indigenous Grp. Retrieved from http://baguiomidlandcourier.com.ph/benguet.asp?mode=%20archives/2017/january/1-22-2017/beng1.txt

The Indigenous – Host Communities Association of Philex Outlying Sitios Inc. (TIHCAPOSI). (1999a). *Articles of Incorporation, TIHCAPOSI.* (unpublished typescript). Tuba, Benguet: TIHCAPOSI Files

The Indigenous – Host Communities Association of Philex Outlying Sitios Inc. (TIHCAPOSI). (1999b). *By-Laws of TIHCAPOSI.* (unpublished typescript). Tuba, Benguet: TIHCAPOSI Files.

Mines and Geosciences Bureau. (2005). Primer on the Social Development Management Program (SDMP). Retrieved from www.mgb.gov.ph/sdmp_frame.htm#primer

NCIP Administrative Order No. 2012–3. (2012, 13 April). The Revised Guidelines on Free and Prior Informed Consent (FPIC) and Related Processes of 2012. Retrieved from http://ncipcar.ph/pdf_files/admin_orders/ncip%20ao%20no%203,%20s%202012%20(fpic).pdf

Northern Philippine Times. (2009, 31 August). Tribal Folk Get P142 M from Philex. 17 February 2011. Retrieved from www.northphiltimes.com/?cat=58&paged=2

PMC (Philex Mining Corporation). (2003). *Philex Mining Corporation Padcal Operations ISO 14001 Certified: 'A Gift of God, a Work of Man'*. Manila: Mimeo.

PMC (Philex Mining Corporation). (2008). Philex Mining Corporation 2007 Annual Report. Retrieved from http://globaldocuments.morningstar.com/documentlibrary/document/27b4efb80e7701e8.msdoc/original

5 Antamok

Launchpad of industrial mining in the Philippines

> You ask us how long might we be doing our pocket mining. I tell you, as long as there is gold in these mountains, and as long as we strive to survive. We mine in order to live. The amount of gold that we are able to dig from the earth is pittance compared to how much the big mining companies have amassed for a very long time. Just like the rice paddies in the *ili* (ancestral village), the holes here are our lifeblood. The owners of the company are not poor, they have fat wealth. We are also citizens of the Philippines whose resources should not be confined to the already rich and powerful.[1]
>
> Celiro Mad-anan (pseudonym)
> A Small-scale gold miner in Antamok

Mining in Itogon demonstrates the convergence of mineral endowment, colonial mining policy, indigenous tenurial systems, industrialization and corporate business goals, and struggles of both traditional communities and settler migrants at securing continuity of their livelihood. This chapter focuses on the 112-year old Benguet Corporation (BC), the country's oldest mining company: its beginnings, growth, decline, and current initiatives and challenges at addressing socio-environmental concerns associated with its open pit in Antamok that closed in 1998. The discussion highlights the interplay of regulatory, economic, cultural, and moral considerations that frame mine closure and rehabilitation processes. A distinct feature of rehabilitation concerns in Antamok pertains to land tenure: BC's mining area is not only tremendously vast, but more than half of the tenements are the company's private lands, i.e., patented claims.

A place of gold past and present

Itogon is a large district of both large-scale industrial mines and small-scale mining sites. Itogon subsumes three places whose very names bespeak of gold mining: Antamok, Acupan and Balatoc. Antamok was the place that the

Spanish referred to in their chronicles as 'Antamog' (Quirante, 1624), which they described as a mountain of many gold sites worked by Igorot people. Antamok was also synonymous and interchangeable with the name *Pancut-cutan* (Antolin, 1789 [Translated by Scott, 1970], p. 192), which denotes 'a place where digging is carried out'. The description derives from both the Ibaloy and Kankanaey root word '*kutkut*', which means 'dig' or 'burrow'. Meanwhile, Balatoc got its name from *balitok*, the local term for 'gold' although some older people explain that *Balitoc* is the name of the god who owns the gold in Itogon. Acupan got its name from the word '*acop*', meaning 'scoop' or 'collect' using both palms. Acupan used to be called *Pangup-cupan* or *Panacupan* which literally translates to 'a place where scooping is done' referring to gathering and heaping up rocks or stones and sand as part of gold extraction and related activities.

Present-day Itogon is one of the 13 municipalities that comprise Benguet which, together with five other provinces, make up the Cordillera Administrative Region (see Figure 5.1). As a municipality, Itogon comprises nine *barangays* whose aggregate population stands at 59,820 (Philippine Statistical Authority, 2015). The barangay is the smallest political unit in the Philippines and serves as 'the primary planning and implementing unit of government policies, plans, programs, projects, and activities in the community' (Republic Act No. 7160, 1991). Antamok is part of Barangay Loacan, while Acupan and Balatoc belong to Barangay Virac.

The municipality of Itogon is the core of the Baguio Mining District. For a long time, nowhere in the Philippines had been a place where large-scale mines, although most of them shut down in the 1980s and 1990s, and 'small-scale' mines were more concentrated than Itogon. This has made Itogon an early destination of thousands of people from many provinces beginning before the Second World War as mining companies recruited workers. While the large-scale mines operated, small-scale mining thrived. Both sectors carried on alongside each other, with the large-scale mines controlling and owning vast areas of land, and the small-scale miners undertaking their operations bordering or located within large-scale claims. With the closure of large-scale mining operations in the area, many displaced employees shifted to small-scale mining. The venture has continued to attract sojourners and visitors who are mainly from the rest of the Cordillera provinces. While people move in and out of Itogon depending on the price of gold, many migrant small-scale miners ultimately choose to reside in the area. Since time immemorial, gold mining has been a significant part of the social life and economy of Itogon's indigenous Ibaloy. Likewise, it is an enterprise which the Kankanaey, who arrived in Itogon beginning the 1920s (Caballero, 1996), are known to be specialists. The Kankanaey have the reputation of being natural geologists for their instinctive ability to identify or detect

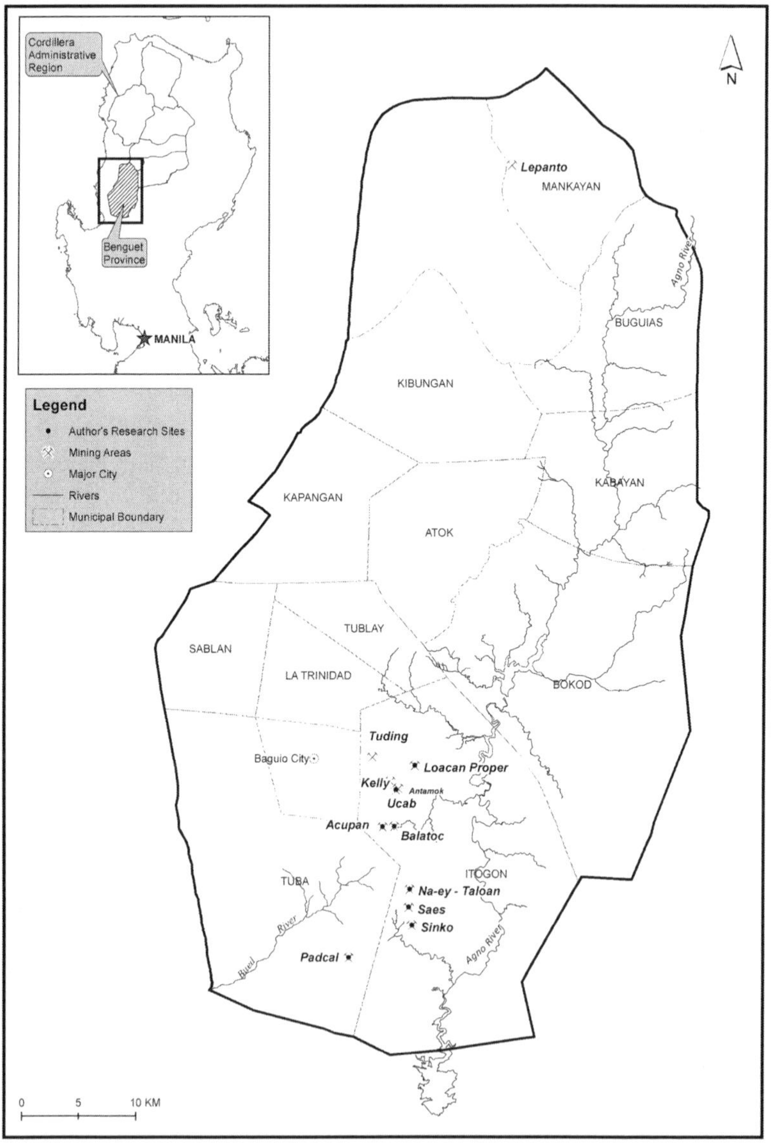

Figure 5.1 Antamok, Acupan, and Balatoc as part of a large mining niche in Itogon, Benguet

(*man-sin-eng*) high-grade gold veins just by 'wetting a rock with their saliva and squinting their eyes'.

The miners in Itogon refer to their venture as 'pocket mining'. The term is attributed to the method in which miners trace gold veins by entering 'pockets', the extremely narrow cracks and clefts of the earth. Benguet miners also call the activity *camote* mining, attributed to digging for camote, the local term for sweet potato (*Ipomea* batata). Other terms associated with pocket mining include *avucay*' ('excavate' or 'dig'), and *usok* ('hole', 'adit', or 'tunnel') (Chaloping, 1991). The term 'small-scale mining'[2] (SSM) is a broader category that related other activities such as panning (*sayo* or *sejo*), milling (*giling*), financing and 'joining' (*jo-win* or *maki-sagaok*) which involves women further re-working tailings. Panning is undertaken by people in downstream communities who rely on gold sediments along waterways that ultimately lead to the Agno River. Meanwhile, milling is the specialisation of a few individuals who charge rent for the use of their ball mill. Some miners do not own ball mills so they bring their ores to mill owners. Depending on arrangements, mill owners may retain the sand washing which can be further worked by panning to recover residual gold. There are individuals who may not be directly involved in mining, but they finance the mining operation of a work group and collect a fraction of the earnings before these are subdivided among the members.

The birth of industrial mining in the Philippines

Modern mining in the Philippines began in Itogon with the establishment of Benguet Corporation. It is the former Benguet Gold Mines that was incorporated on 24 June 1903, and registered as *sociedad anomina* by three Americans: Metcalf A. Clarke, Nelson Peterson, and Henry C. Clyde. In 1906, the company commenced mining and milling operations in Antamok and changed its name to Benguet Consolidated Mining Company (BCMC). The Antamok mine was the first underground gold operation commencing in 1907 and continued until the outbreak of the Second World War, except in 1909 and 1911 when two typhoons damaged the company's mill facilities (De Jesus, 1978). The mine resumed in 1948 and achieved the highest postwar productions from 1951–1973 with annual yields ranging from 87,284–129,246 ounces of gold (BC, 2010a).

The rich mineral lands of Antamok were once under the control[3] of Jose Fianza, a *baknang*,[4] a person of the upper class rank in Ibaloy society who served as the sole buyer of all gold in Itogon, Bokod, and Kabayan or the entire lower Agno Valley (Chaloping, 1991). Capitalists bought Fianza's mining claims and later developed the property. Acupan, which is located several kilometres south of Antamok, was similarly a rich area of old lode mines acquired and developed by Major Harry Howland and Clinton Huebert to become the Acupan Mining Company in 1909. During this time, there were scores of exploring, grubstaking, and 'acquiring' such that mining claims in Itogon totaled 80 (Lopez, 1992). There was a 'furor of speculation' and feverish excitement towards the 1930s as mining machinery and large-scale capital for the industry flowed into the Philippines (The Saturday Evening Post, 27 March 1937).

Pioneering pursuits, unparalleled growth, and decline

BCMC's becoming the country's largest gold producer was attributed to the entrepreneurial skills of John W. Haussermann, a second lieutenant of the 20th Kansas Volunteers of the U.S. Army, who took over as company president. As early as 1916, BCMC shareholders received their first dividend of PhP100,000 or US$50,000. In 1920, the company produced over PhP1 million worth of gold-bullion (BC, 2005b). While serving as President of BCMC, Haussermann also offered financial and technical support to other companies with promising mineral properties within the Baguio-Itogon district. His provision of initial capitalization for the development of other mines led to the enlargement of his enterprise. In 1927, BCMC acquired Acupan Mining and Balatoc Mining (BC, 2005b; Lopez, 1992). The Second World War wrought havoc on BCMC's facilities but the company began rehabilitation in 1946 and resumed operations in 1947.

On 18 June 1956, BCMC was renamed Benguet Consolidated Incorporated (BCI), launching the company's remarkable expansions including diversifying into engineering and industrial construction, agriculture, shipping, banking and finance, real estate, and forestry-based ventures. By 1965, BCI was the largest gold producer and biggest gold cyanidation mill operator in the Philippines and the Far East. Its gold output in 1970 accounted for about 50 percent of the entire gold production of the Philippines (BC, 2005b). The ascent of Ferdinand Marcos to the presidency had significant implications for the company's ownership structure. Under President Marcos, legislation limited foreign ownership of the country's mining companies to just 40 percent. This resulted in BCI coming under the control of the Romualdez family, primarily Benjamin Romualdez, brother of Imelda Marcos (Ramos, 1988). The Marcoses had been buying into shares of what was then BCMC. In this way, the company was being turned over to the Romualdez family, a scheme by which Marcos made concessions to his wife (Rempel, 1993, p. 92; Seagrave, 1988, p. 228).

In 1980, BCI renamed itself Benguet Corporation (BC), its present name, and further grew. It enlarged its mill in Balatoc to process lower-grade gold ore and operated several mines simultaneously. It was operating the Baguio Antamok Gold Operation (BAGO), and the Benguet Gold Operation which subsumes Atok, Acupan, Sangilo, and Kelly mines. Simultaneously, it was operating three more major mines outside Benguet Province (BC, 2014), namely Paracale Gold Operation (Camarines Norte), Dizon Copper Operation (Zambales), and Masinloc Chromite Operation (Zambales). The company also started in the mid-1980s investing considerably in its original Antamok mine with the aim of further increasing its production by extracting higher-grade ore deposits (Ongpin, 1983). BC has not restricted its aspiration of growing in the Philippines only. As early as 1949, the company became the first in the Philippines to be listed at the New York Stock Exchange. Two decades later, the company established Benguet International, S.A. to advance into the international market. In 1988, BC acquired a Philippine-owned Hong Kong-based company and named it BenguetCorp International, Ltd. to pursue mining opportunities at an international scale (BC, 2005a).

During the 1990s, BC's operations were gravely affected by natural calamities such as the Baguio earthquake in July 1990, the eruption of Mount Pinatubo in 1991, and countless typhoons. Low metal prices at the international market and the 1997 Asian currency and economic crisis compounded the situation. These brought BC's several operations to a halt: the Benguet Gold Operation in 1992, the Paracale Gold Operation in 1993, and the BAGO in 1998. The company also decided to sell its remaining interest in Dizon Copper Operation in 1997 when its facilities damaged by the 1991 Mount Pinatubo volcanic eruption rendered rehabilitation too costly. The company also turned over Masinloc Chromite Operation to the claim owner in July 2007 as the operating contract expired.

The Antamok Gold Project

The aerial expanse of the Antamok mine is approximately 5.24 square kilometres[5] and comprises 74 mining claims of which 3.28 square kilometres are patented.[6] The mineral claims are either owned or controlled by BC, which the company has acquired through several generations of laws, i.e., the Philippine Bill of 1902,[7] Commonwealth Act No. 137,[8] Presidential Decree No. 1214[9], and Presidential Decree No. 463[10] (BC, 2010a, p. 10) under the American regime and under the administration of President Ferdinand Marcos. The company's proprietary rights over its patented claims include surface rights (BC, 2017).

After 86 years of underground operation, the Antamok mine's high-grade gold ores were depleted and only low grade ores occurring at or near the surface remain. Thus, BC launched in 1989 the Grand Antamok Project, renamed as the Antamok Gold Project (AGP), and finally BAGO. It involved open pit bulk mining that targeted the remnants of the wide veins that had been extracted over many years (BC, 2010a). The Antamok operation stands on 3.28 square kilometres of BC's patented mining claims.

From the time the AGP was just taking off in 1989, it had faced intense opposition from the communities within and near the project area. The identified total coverage for the company's open pit operations include not only Antamok's minesite camp – areas of cottages and bunkhouses occupied by wage-earning BC workers and their families – but also lands outside the perimeters of BC's industrial and residential sites. These include burial grounds, water sources, backyard gardens, swidden farms, and mining sites but most of these lands lie within patented claims of BC. The work sites for pocket mining are within Ucab and Loacan, two barangays of Itogon that were most immediately affected by BC's open pit operations in the 'Keystone' and '440' veins, which are both located within the company's patented mining claims.

Two case studies on the Antamok open pit mining, by Carino (1993b) and by Gimenez (1991), describe the communities' aversion to the project as directed against what they considered destruction of their properties (particularly ancestral lands and houses), livelihood and income sources, their environment, and their way of life. As the project pushed through, company bulldozers and other heavy equipment razed and flattened lands, including mining sites, farms, and water sources (Carino, 1993a). Various non-government organisations (NGOs) which were based in Baguio helped in mobilising barricades and marches. These include the Cordillera Environmental Concerns Committee, the Cordillera Peoples Alliance, the Cordillera Resource Center for Indigenous Peoples Rights, and the Mining Communities Development Center (MGDS-CAR, 1992). Organised community groups include the United Concerned Citizens of Ucab, Loacan Itogon Pocket Miners Association, Loacan Landowners Association, and Antamok Miners Association, *Nagkaykaysa nga Umili ti Gumatdang* (United Citizens of Gumatdang), and Dalicno Traditional Miners' Association (MGDS-CAR, 1992). The affected communities established the *Timpuyog Daguiti Umili Iti Itogon* ('Alliance of the Citizens of Itogon') to oppose the project (Carino, 1993a) on the claim that BC did not comply with Presidential Decree 984 (Pollution Control Law) and Presidential Decree 1586 (Environmental Impact Assessment [EIA] Law). The heightened pickets, marches, rallies, and barricades involved incidents of arrest and detention of community members by the police (Carino, 1993b).

When full-scale open pit operations recommenced in January 1990, the company asserted its legal rights in pursuing its project. Correspondingly, the people met such resolve with equal decisiveness by setting up barricades. A regional trial court issued in March 1990 an order directing the communities to lift their barricades. In May 1990, as the positions of the opposing parties stiffened, the DENR coordinated and organised dialogues and public hearings for the contending parties. The DENR proposed a compromise scheme to solve the standoff: allow the open pit mining activities to resume in return for indemnities from the company (Carino, 1993a). The compromise was formalised in a Memorandum of Agreement (LIPMA File, 1990) signed on 7 June 1990 by BC, leaders of small-scale miners associations, landowners of Loacan, and the DENR. The MOA provided for the dismantling of community barricades by the protesters. It also stipulated that BC will withdraw the court suit it filed against 27 leaders of the protests, pay compensation to affected people, and provide opportunities for labour contracts to residents of Loacan.

With the MOA signed by heads of the parties involved, the DENR granted BC on 27 June 1990 the much-sought-after environmental compliance certificate (ECC) that signalled the project to operate. Only a few weeks following the signing of the MOA, local leaders expressed anew their opposition to the AGP, alleging co-optation and misrepresentation against the individuals who signed the MOA. In the context of the perceived duplicity by their leaders, the protesters established new organisations at the barangay and *sitio* (sub-village) levels to form in August 1992 a municipality-wide coalition: the Itogon Inter-Barangay Alliance (IBAA) (Carino, 1993a). The community opposition led by IBAA continued. In effect, BC faced difficulties in proceeding with its operations because it could not secure right of way and access to critical areas for mining. In addition, the gold prices at the international market plunged. BC eventually suspended the project in April 1998, and the Antamok open pit mine was put on care and maintenance status (BC, 2010a).

An environmental policy borne out of Antamok

The resistance to the Antamok open pit mining created an important juncture in the history of mining policy in the Philippines. A leading community leader, the barangay captain of Ucab at the time the opposition was being waged, sees the Antamok experience as having raised clearly and strongly the point that people, not only the environment, must unquestionably be indemnified when a mining operation causes damages. He observed:

> It was the experience in Itogon that brought about the idea of environmental guarantee fund (EGF) in our laws. The problems that the people raised concerning compensation for damages to their communities and their surroundings effectively confronted the DENR. It was real! Before

> their very eyes, DENR officers witnessed people fighting for their livelihood and homes. DENR had to seriously do something about compensating people. So, they introduced the EGF. If not because of the fight put up against the open pit project, I do not think there would be EGF.[11]

The Philippine Environmental Impact Statement (EIS) system incorporated the requirement for EGF through the issuance of Department Administrative Order (DAO) No. 21 (1992) or 'DAO 21', amending the Revised Implementing Rules and Regulations of the EIA Law.[12] This policy made the EGF part of the monetary fees collected from industries that have potential to cause catastrophic damage to the environment or pose significant public risk. The fund is allocated for rehabilitation and compensation for damages and other activities such as 'research works, environmental education, and programs to promote social equity'. In the few years that followed since DAO 21 was issued, the DENR has strengthened the EIS system by incorporating new requirements in response to changing realities and the growing environmental consciousness of the public. Thus, the DENR issued a new administrative order, DAO 37 (1996),[13] which incorporate into the EIS system specific elements for more comprehensive assessment of proposed projects. Two of these elements are public participation and social acceptability to focus on the role and concerns of communities in the EIA process. DAO 37 further prescribed the mandatory inclusion of EGF, and officially established the EIA Review Committee, a pool of certified experts from which specialists of different fields are drawn to appraise environmental impact statements of project proponents.[14]

The DENR considers the AGP as the first[15] of three major events in the Philippines that sparked vehement NGO opposition to mining (Cabalda, Banaag, Tidalgo, & Garces, 2002). In 2015, the DENR issued DAO 02 (DAO 2015-02, 2015)[16] to synchronise the EIA Law and the Mining Act of 1995 particularly the intersecting provisions on environmental management and social development in mining projects. The policy requires all proposed mining projects to set up Contingent Liability and Rehabilitation Funds, Environmental Trust Funds, and funds for Social Development Management Programs, in lieu of the EGFs and the Environmental Monitoring Funds. The purpose is 'to defray the cost of monitoring, compensation for damages, social development, progressive rehabilitation, and final mine rehabilitation/decommissioning.'

Recollections on the Antamok Gold Project

The AGP is an event that is etched in people's memories. An official of the Benguet Provincial Government, who was the mayor of Itogon municipality at the time the opposition was mounting against the AGP, regards the project as instrumental in bringing out the defiance of the generally gentle and quiet Ibaloy and Kankanaey of Itogon. He added: 'the company had the legal rights to pursue its project, but the issue became a moral one – people losing their homes and livelihood.'[17] To a 47-year-old resident in nearby Balatoc, the open pit in Antamok is an unpleasant story to be told and retold and will not go away. He adds: 'the huge yawning hole as it is now, which was once an imposing mountain, could have been bigger if the people of Itogon did not lift any hand to fight.'[18]

Meanwhile, to a nationalist advocacy organisation like the Cordillera Peoples' Alliance (CPA), which actively supported the communities in Itogon, the protests signify people's determined actions that effectively constrained BC's bulk mining to a single open cast operation in Antamok which turned out to have drained the company's coffers such that the operation had to cease, making it five years short of the project's supposed 12- or 13-year mine life. The CPA, describing its 'Past battles with Benguet Corpn', notes that with seven barangays participating in opposing the AGP by holding barricades that lasted for three years, many communities in Itogon were spared major devastation (CPA, 2004).

The opposition to the AGP is a significant juncture in crystallising a shared local nationalism among Itogon residents. The recollections about the 'struggle' could have been cast solely with deep gratitude to the NGOs that helped them. However, a few individuals describe one aspect of NGO assistance as 'regrettable'. Among a group of traditional pocket miners in Loacan, a former leader in the mass protest against BC, recalls:

> We were opposing the open pit project in Antamok during those years. The 'En *Jee Oos*' were very helpful organising us and leading the barricades. However, I disagreed with their move of collecting jeepney-loads of street children from Baguio to bring to the barricade site. From hindsight, it became clear to us that some 'En Jee Oos' had used (*inusar*) people – children, and even ourselves, the residents. . . . When the arrests took place, the organisers and leaders were gone, although they had made strong issues out of our arrest. One of the leaders has now made a name for himself out of the concerns we were raising then. But he is *persona non-grata* in Itogon now. We do not want to see him ever again.[19]

The *kapitan* (local chief executive) of Barangay Loacan has one vivid memory about the AGP. He recalls:

> The event was a real trauma to us – myself, my family, my neighbours, my community, all of us in Loacan. That was more than 30 years ago but no one can erase it. It did happen. It was traumatic because we saw

> lands being bulldozed. We heard drilling and blasting. We people in Loacan did not want confrontation so we were grateful that the dialogues organised by the DENR eventually brought us and the company to a compromise. We signed the MOA. [Here, you may read it]. We continue to hold onto what we signed in 1990 – that BC will allow us, the Loacan Itogon Pocket Miners Association, to move back to 440 Vein and the Camote Vein to continue our mining activities for as long as there is viable gold, and as long as we want.[20]

Faced with developments associated with rehabilitation imperatives for BC to undertake in the area, Loacan residents – through the Loacan Pocket Miners Association (LIPMA) and the barangay council – are taking moves to ensure their mining activities will carry on. The kapitan continues:

> *Ti balitok ditoy dagami isu ti nagtaudan mi nga Ibaloy ditoy Loacan* (The gold underneath our lands is one from which we the Ibaloy people of Loacan sprung). Our way of mining is the traditional way – we invoke the spirits of our ancestors through our ritual practices. We perform ceremonies and make offering by butchering an animal at least once every year, lest loss of life takes place. It is Kabunian who owns the gold, and it is no joke if we veer away from the values our ancestors have always observed. Our ancestors taught us not to be gripped by greed. Accidents happen when we forget that gold is for the community's sustenance. Avarice must not take place. May disunity not happen in the community. We will be doing our best to have a Minahang Bayan. We will negotiate with the company on this because the location of their intended waste facility is something we must talk about.

In Antamok where the open pit project ceased in 1998, many displaced BC employees and their families left. However, an equally huge population continued residing in the area and resorted to pocket mining as an economic activity, adding to the already increasing numbers of pocket miners. Some of the former BC employees and their children built their houses adjacent the open pit area while others were able to settle in Surong and Luneta, two dense communities located with in the company's extensive patented claims. They were spared from the bulldozing and blasting activities of BC during the late 1980s as the company implemented its open pit operations in the 440 Vein and Keystone Vein sites, which are also within BC's patented lands.

At present, the pit and its vicinities have been the work sites of countless pocket miners with many of them members of, initially, at least three organisations – the Luneta Miners' Association, Antamok-Saldin Miners'

Association, and Loacan Itogon Pocket Miners' Association. In the last few years, however, more pocket miners' associations emerged whose memberships include mainly migrants from various provinces of the Cordillera, particularly Mountain Province and Ifugao. The Ibaloy pocket miners of Itogon have reasons to feel outnumbered. Some remark that they feel uneasy about their homeland being invaded by non-Ibaloy whose numbers add to the increasing competitors for Itogon's gold.

Rehabilitating the open pit

The Antamok open pit mine had operated for about seven years until it shut down in April 1998. Most of the infrastructure of the original mine were demolished when the mine converted from underground to open pit operation. The underground opening below L-700 underneath the bottom of the pit has been 'flooded and rendered inaccessible' (BC, 2010a, p. 22).

After the AGP, the huge crater has not been backfilled. BC refers to it as the '440 West Open Pit' while others call it simply '440' or 'open pit'. The open pit has continually gathered silt from ordinary erosions and major landslides especially during the rainy season, and from continuous excavations by pocket miners who have made the pit and its vicinity their workplace. Hundreds of them also enter Antamok's old abandoned underground mine leading to 'Level 70' – or 'L70' – often endangering themselves. In previous years, accidents have resulted in loss of lives (Manila Bulletin, 2010; The Philippine Online Chronicle, 2008). A bit farther across the 440 pit is a site called 'Camote Vein', another area where scores of pocket miners work. Numerous mine holes and makeshift shelters strew all sides and walls of the 440 pit, Camote Vein, and L70 (BC, 2017). BC has not been able to secure its properties because its personnel are outnumbered. Efforts of the MGB, provincial and municipal governments, and BC to regulate them have not succeeded. Legal battles filed by BC against some pocket miners have continued to be heard in the courts until the present.[21]

Part of the effort at regulating the illegal entry and operation of pocket miners was BC's offer to make them partners in a contract venture just like the Acupan Contract Mining Project (Chapter 6). BC launched a pilot test on establishing the Antamok Contract Mining Project (AnCMP) in 2007 which involved requiring an entrance fee of PhP15.00 for every pocket miner each day to the underground workings of L70. Pocket miners are grouped into cooperatives or *kumpanya* to work in designated areas. The venture lasted for only three years (2002–2005) because the pocket miners found the entrance fee too burdensome. BC also presented and discussed with the pocket miners a production sharing scheme but the parties failed to come into an agreement. In June 2008, the Mines and Geosciences Bureau-Cordilleria

Administrive Region (MGB-CAR) suspended the AnCMP as part of a cease-and-desist order (BC, 2010b). Subsequently, DENR issued a mining moratorium in the Antamok area in January 2009.

Bulk water supply project

As a rehabilitation strategy for the pit, BC considered converting it into a water reservoir (BC, 2004). In August 2005, the Baguio Water District (BWD) awarded to BC the Bulk Water Supply Project, after having won the project bidding held in 2003. The project aims to supply 50,000 cubic metres per day of potable water to Baguio City supposedly by mid-2008. BC formally engaged Deutche Bank as financial advisor to assist in the financing of the project which was estimated to cost US$60 million (BC, 2006). The water project did not elude opposition by residents of Itogon and Baguio. The main apprehension of the people in Itogon is the possible depletion of current water sources considering the 50,000 cubic metre capacity which BWD requires the company to deliver to Baguio daily, should the project materialise.

Critics of the company and the mining industry raise questions about resource rights of residents against a business entity's plan to commoditise water. They also raise the need for meaningful dialogue between residents and BC. According to the company, it has conducted its required consultation with the public and local government units about the project, and the results show support from the municipality officials. Meanwhile, the Baguio consumers' key concern is about the quality of the water which would include run-off from highly mineralised lands, and the open pit is alleged to contain noxious substances (Northern Dispatch, 2005, 14 August). However, BC explains that the Bulk Water Project incorporates a water treatment plant to ensure water is potable before distribution. In September 2007, BWD terminated the contract negotiation with BC on the project. BC requested the BWD conduct a public hearing on the issues of contract provision, which the BWD deemed premature, and denied the former's request for reconsideration. In response, BC filed a case against BWD (BC, 2008, p. 32), which still remains unresolved at the Baguio Regional Trial Court until the present.[22]

With the action of BWD, which BC regards as arbitrary, the Bulk Water Project takes a back seat. The company is now considering developing the open pit as an ecotourism venture which will incorporate a water park with facilities for sports and recreation as well as wellness and leisure activities to include forest trekking and hiking. The plan involves landscaping the area of 6.8 square kilometres with mixed reforestation species of various plants for slope and buffer zone protection. While BC is planning for this ecotourism project, it does not abandon the prospect of the proposed Bulk water Project (BC, 2017).

Engineered sanitary landfill

Meanwhile, it emerged rather opportune for BC that the city government of Baguio has been in search for a suitable location for a sanitary landfill to address its perennial solid waste problem. Under existing legislation, particularly Republic Act 9003 which was passed in 2001, all local government units, such as the city of Baguio and the Municipality of Itogon, must establish an ecological solid waste management program. In 2008, BC offered its open pit for consideration by the city Government of Baguio to use as an engineered sanitary landfill (ESL).

BC signed an MOA with the mayor of Itogon and the mayor of Baguio on May 2016 and January 2017 on the proposed ESL project. However, the project will no longer be located at the open pit but at the Camote Pit, which incidentally is a current mining site of the Loacan Pocket Miners Association (LIPMA). BC is currently conducting a review on the boundaries of the ESL to address the concerns of the pocket miners. The ESL project also involves Goldrich, a private developer whose services BC has procured to prepare the feasibility study and detailed engineering design. The ESL will be an initial component to the waste energy project which the developer will construct at the former Antamok Mill site (BC, 2017).

Minahang Bayan

Antamok as a mining area has been overrun by pocket miners whose huge numbers – reaching as many as 5,000 (with each miners' association having 1,000–1,500 members) – are too arduous for BC to prevent from entering the company's mineral properties and claims. On several occasions, the company management had to secure cease-and-desist-orders from MGB-CAR for the Philippine National Police to enforce. Considering that the AnCMP did not work, BC is willing to establish and locate an area for *Minahang Bayan* where pocket miners can pursue their livelihood.

The Minahang Bayan area proposed by BC is 0.52 square kilometres for which the company submitted a proposal to MGB-Cordillera Administrative Region (CAR) for review and validation.[23] If granted, the Minahang Bayan in Antamok will become the first in the region, and the sixth in the country. The officials of Barangay Loacan, together with members of different pocket miners' associations in Itogon, met with the DENR Secretary Gina Lopez during her visit in Loacan on 3 February 2017. The secretary announced her support for the establishment of a Minahang Bayan.

The Liang tailings pond

The pond located upstream of the convergence of Antamok and Liang Rivers, which has impounded tailings of about 9 million tonnes, was part of BAGO. As an alternative land use, BC has been considering to further recover gold from the tailings pond. Thus, the company employed in 2012 the services of Envitech, an environmental management consultancy firm, to prepare a feasibility study for the purpose. In the same year, BC applied at the MGB for a mineral processing permit to process the tailings and recover gold values to produce concentrates (BC, 2017).

Incidentally, the Municipal Council of Itogon prepared a resolution[24] (MCI, 2016, 15 November) calling on BC and concerned agencies to implement actions concerning the leakage of the Liang Tailings Pond on 27–28 October 2016 as well as the 'existence of surface tension cracks within the vicinity of the Antamok Open Pit'. The resolution lists 14 actions the municipal body wants carried out. One of these is for BC in coordination with MGB-CAR and Environmental Management Bureau-CAR 'to validate and make public the actual cause of the tension cracks and the breach/failure of the Liang Tailings Dam' towards implementing immediate remedial measures. The resolution also asks 'final full rehabilitation of the Antamok Open Pit and talings dam at the company's cost' (Municipal Council of Itogon, 2016, p. 3). Two weeks before the resolution was issued, BC submitted to the Philippine Stock Exchange Commission its press release (PSEC, 2016) noting: 'an intermittent leak of mine tailings was discovered following regular inspection conducted by the company after heavy rains brought about by Typhoon Lawin.' The report noted that the company 'has mobilized its workforce and equipment to contain the leak and has initiated cleanup operations in coordination with MGB, EMB, local communities and LGUs.' The two separate information demonstrate that BC and the Itogon Municipal Council are not on the same wavelength.

Reliving old Antamok

As early as the 1930s, the Antamok minesite camp was already a compact settlement of cottages and bunkhouses whose residents included many rural dwellers who left their provinces the very first time in their lives. Thousands of mine workers and their families arrived in Antamok which would become their adopted homeland. Language and ethnolinguistic affiliation served as 'the most conspicuous organisational principles' in the workplace according to De Jesus (1978, p. 43), who also observed that the Antamok mine had 2,000 workers in the 1930s including 'Igorots', 'Pangasinenses' who make up another one-third, and 'a mix of Ilocanos, Pampangos, Tagalogs, Visayans, Bikolanos and mestizos' making up the rest.

The minesite camps in Benguet are the forerunners of modern-day 'overseas' as sources of cash, just as the mine workers in Benguet are the old version of today's overseas Filipino workers or 'OFWs'. Part of mine workers' income was spent for goods that were sent to their hometowns. In many Cordillera villages, the traditional thatched roofs of houses began changing into galvanised iron sheets in the 1960s towards the 1970s mostly with incomes derived from the mines. Farm tools such as shovels, digging forks and grab hoes, rubber boots, and even miners' skull guard hats have found their way in some homes in the villages. By the 1980s, the workforce in the Antamok mine has undergone at least four to five generations, with high literacy and producing a considerable number of successful university graduates. Many professionals in the Cordillera region acquired their primary and secondary education in company-built, -administered and -subsidised schools (Habana, 2001).

In 1988, BC's Benguet Gold operations had a total workforce of 5,333 including 288 managers (BC, 1988, p. 10). This number began shrinking with the retrenchment of 856 employees in 1989 as part of the gold operational restructuring program (BC, 1989, p. 7). Most of BC's operational infrastructure for both mine operation and workers' accommodation were torn down during the AGP. Following the company's suspension of operations in 1992, BC had to dispose of materials its personnel could recover from buildings, machinery, and other assets.

With the AGP, Antamok has changed forever. Nonetheless, countless men and women who were born, grew up, worked, and resided in Antamok have had the common experience of having *lived* in the minesite camp. Many mining engineers and other mine workers look back to Antamok as their training ground and career-niche. Numerous teachers, lawyers, doctors, and members of other professions, countless of whom now live overseas, look back to Antamok with nostalgia despite the dissolution of the minesite camp they all knew. The appellations *Taga-Antamok* ('of'/'from Antamok'), *Taga-Minas* ('of'/ from the Mines') and *Antamoker* are socio-cultural-spatial markers that former residents assume in keeping themselves identified with, and the yearning to re-live, the place they consider a significant part of their life.

In social media at present, several Facebook pages are devoted to the purpose of keeping connectedness among those who have or had lived in Antamok. For instance, 'Forever Antamok' which is a public group of 1,935 members bears the description 'private space for our family to share photos, videos, and notes.' The page posts announcements to former residents of Antamok about events such as reunions in respective countries, e.g., London, Taiwan, and specific states in the USA; deaths and expressions of condolences from old neighbours; as well as parties among 'Antamokers' during special occasions in the Baguio-Benguet district or overseas. A larger group, *Anak BCI* ('Child of BCI'), is a closed group of 2,955 members that describes itself as 'place for friends and families who used to live in Acupan, Balatoc and Antamok to congregate and maintain a sense of community through the internet.' Similarly, the webpage features many photos of the old BCI minesite camps and new videos of the current landscape including the Antamok open pit. Anak BCI has a dozen chapters spread out in Australia, the Middle East, Indonesia, and Hong Kong.

Notes

1 Author's interview, February 2017, Luneta, Antamok.
2 Republic Act No. 7076 (1991), also known as the People's Small-Scale Mining Act, defines small-scale mining as 'minimum activities which rely heavily on manual labor using simple implements and methods, and which do not use explosives or heavy mining equipment'. This definition is debatable as there are groups whose methods involve advanced technology and use of explosives.
3 'Control' over the land was usufruct, which means the person 'in control' has access to the mining area/land for as long as he *uses* it or works in the area.
4 *Baknang* literally means 'rich', a leader in the Ibaloy region owing to his economic and political influence which is based mainly on wealth that included cattle, pasture lands, and mining sites.
5 Company data which are in hectares have been, in this book, converted into square kilometres.
6 A patented mining claim is private land.
7 Under the Philippine Bill of 1902, the holder of a mineral claim is entitled to all minerals which may lie within his claim.
8 Commonwealth Act No. 137, otherwise known as the Mining Act of 1936, provided that any existing right, grant, lease, or concession at the time of the inauguration of the government established under the 1935 constitution shall be respected, such as patented and patentable mining claims.
9 Presidential Decree 1214 (1977) required all locators under the Philippine Bill of 1902 to secure mining lease contracts under the provisions of Presidential Decree No. 463 involving their patentable mining claims; otherwise, all the rights of locators to their patentable claims will be forfeited and the land reverted back to the public domain.
10 Presidential Decree (PD) No. 463, promulgated in 1974, compelled claimants to actively develop their claim to prevent speculation in mining claims.
11 Author's interview, October 2005, Baguio.
12 Presidential Decree 1586, which was issued in 1978, requiring the compulsory submission of environmental impact assessment of environmentally critical projects, mining included.
13 Revising DAO 21 to further strengthen the implementation of the EIS.
14 I was an accredited and certified EIA Reviewer for EMB. As member of the EIA Review Committee during 1997–1998, my task involved assessing the social acceptability of mainly mining and energy projects.
15 The second two events are the enactment of the Mining Act of 1995 and the Marcopper spillage in 1996 (Cabalda et al., 2002).

16 'Harmonization of the Implementation of the Philippine Environmental Impact Statement System and the Philippine Mining Act of 1995 in Relation to Mining Projects'.
17 Author's interview, April 2005, La Trinidad, Benguet.
18 Author's interview, March 2005, Balatoc, Virac.
19 Author's interview, October 2004, Baguio.
20 Author's interview, February 2017, Loacan Proper, Loacan.
21 Personal communication with BC's chief of staff to the officer-in-charge. April 2017.
22 Author's interview, February 2017, Baguio.
23 Personal communication with BC's chief of staff to the officer-in-charge. April 2017.
24 The author received a copy of the resolution on February 2017.

References

Antolin, F., & Scott, W. H. (1789, translated by W.H. Scott in 1970). *Notices of the Pagan Igorots in 1789 in the Interior of the Islands of Manila, of Their Gold and Copper Mines and Their Commerce and Their Various Expeditions, Endeavours and Expenditures Made for Their Discovery and Pacification*. Manila: University of Santo Tomas Press, c1988.

BC (Benguet Corporation). (1988). *Benguet Corporation and Subsidiaries 85th Year Report*. Ortigas Center, Pasig, Philippines: Benguet Corporation.

BC (Benguet Corporation). (1989). *Benguet Corporation and Subsidiaries Annual Report 1989*. Ortigas Center, Pasig, Philippines: Benguet Corporation.

BC (Benguet Corporation). (2004). 2004 First Quarter Report Benguet Corporation and Subsuidiaries. Retrieved from www.pse.org.ph/html/Listedcompanies/pdf/BC_A17Q1of1stQtr2004.pdf

BC (Benguet Corporation). (2005a). Benguet Corporation 2004 Annual Report (SEC Form 17-A). Retrieved from www.benguetcorp.com/html/BC%202004%20Annual%20Report.pdf

BC (Benguet Corporation). (2005b). *A Century of Facts: Benguet Corporation the Natural Resource Company (Compact Disk)*. Makati Philippines: Benguet Corporation.

BC (Benguet Corporation). (2006, 21 August). Benguet Corporation 2005 Annual Report (SEC Form 17-A). Retrieved from www.pse.com.ph/html/ListedCompanies/pdf/2006/BCB_17A_Dec2005.pdf

BC (Benguet Corporation). (2008). Benguet Corporation 2007 Annual Report (SEC Form 17-A). Retrieved from www.benguetcorp.com/html/BC%202006%20Annual%20Report~SEC%20Form%2017A.pdf

BC (Benguet Corporation). (2010a). The Antamok Gold Mine: A Summary of Its Remaining Potential. Retrieved from www.benguetcorp.com/index.php?option=com_content&view=article&id=181&Itemid=95

BC (Benguet Corporation). (2010b). Benguet Corporation 2009 Annual Report (SEC Form 17-A). Retrieved from www.benguetcorp.com/html/BC%202009%20Annual%20Report~SEC%20Form%2017A.pdf

BC (Benguet Corporation). (2014). Benguet Corporation Annual Report 2013 (SEC Form 17-A). Retrieved from http://benguetcorpcom.azurewebsites.net/wp-content/uploads/2014/07/BC-2013-Annual-Report-SEC-Form-17A.pdf.

BC (Benguet Corporation). (2017). *Amended Final Mine Rehabilitation and/or Decommissioning Plan (FMR/DP)*. Baguio City: Mines and Geosciences Bureau Regional Office – Cordillera Administrative Region.

Cabalda, M. V., Banaag, M. A., Tidalgo, P.N.T., & Garces, R. B. (2002). Sustainable Development in the Philippine Minerals Industry: A Baseline Study. Retrieved from http://pubs.iied.org/pdfs/G00614.pdf

Caballero, E. J. (1996). *Gold from the Gods: Traditional Small-Scale Miners in the Philippines*. Quezon City: Giraffe Books.

Carino, J. K. (1993a). *Itogon Inter-Barangay Alliance*. Baguio City: Cordillera Studies Center, University of the Philippines College Baguio.

Carino, J. K. (1993b). *The Mass Movement against Open-Pit Mining in the Cordillera*. Baguio City: Cordillera Studies Center, University of the Philippines College Baguio.

Chaloping, M. M. (1991). *Towards Sustainability of Household Well-Being: Labor Allocation Between Rice Farming and Goldpanning in Dalupirip, Itogon*. (Master of Arts in Social and Development Studies). Baguio: University of the Philippines.

Cordillera Peoples Alliance. (2004, 31 July). Apit Tako on Struggle for Water Resources: Water War? Retrieved from www.cpaphils.org/campaigns/water-war_4731.htm

De Jesus, E. C. (1978). *Benguet Consolidated, Inc. 1903–19878: A Brief History*. Makati City, Philippines: Benguet Consolidated, Inc.

DENR Administrative Order No. 1992–21. (1992). Amending the Revised Rules and Regulations Implementing P.D. 1586 (Environmental Impact Statement System). Retrieved from http://policy.denr.gov.ph/1992/ENV_DAO_92-21.pdf

DENR Administrative Order No. 1996–37. (1996, 6 June). Revising DENR Administrative Order No. 21, Series of 1992, to Further Strengthen the Implementation of the Environmental Impact Statement (EIS) System. Retrieved from http://emb.gov.ph/wp-content/uploads/2016/06/DAO-1996-37.pdf

DENR Administrative Order No. 2015–02. (2015, 10 March). Harmonization of the Implementation of the Philippine Environmental Impact Statement Sytem and the Philippine Mining Act of 1995 in Relation to Mining Projects. Retrieved from http://server2.denr.gov.ph/uploads/rmdd/dao-2015-02.pdf

Gimenez, L. A. (1991). *Going against the Giant: The Pocket Miners of Itogon and Benguet Corporation's Antamok Gold Project*. Baguio City: Cordillera Resource Center for Indigenous Peoples' Rights.

Habana, O. M. (2001). Gold mining in Benguet: 1900–1941. *Philippine Studies*, *49*(1), 3–41.

LIPMA (Loacan Itogon Pocket Miners Association) File. (1990). *Memorandum of Agreement Made and Entered into on 7th June 1990 by Benguet Corporation, Igorot Small-Scale and Panners Foundation, Loacan Itogon Pocket Miners Association, Luneta Miners Association, Landowners of Loacan Itogon Benguet, and the Department of Environment and Natural Resources-Cordillera Administrative Region*. Loacan, Itogon, Benguet: LIPMA Records.

Lopez, S. P. (1992). *Isles of Gold: A History of Mining in the Philippines*. Singapore: Oxford University Press.

Manila Bulletin. (2010, 13 April). Governmentt Pinpoints Tragedy for Sealing Mine Tunnels. Retrieved from www.mb.com.ph/node/252478/

MGDS-CAR (Mines and Geo-Sciences Development Services – Cordillera Administrative Region). (1992). *Mineral Resource Development in the Cordillera Administrative Region*. Baguio City: MGDS.

MCI (Municipal Council of Itogon). (2016, 15 November). *Resolution Number 451, Series of 2016: Resolution Expressing Major Concerns and Required Actions in Relation to the Benguet Corporation (BC) Antamok Open Pit and Tailings Dam Incident, This Municipality*. Poblacion, Itogon: Municipal Council Files.

Northern Dispatch. (2005, 14 August). BC's Water Unsafe – Itogon Elders. Retrieved from www.nordis.net/news/2005/ndw050814/ndw050814_05bwsp.htm

Ongpin, J. V. (1983). The risk management strategy in mining (or Can Benguet Corporation survive to celebrate its centennial in 2003?). *Canadian Institute of Mining, Metallurgy and Petroleum Bulletin, 76*(858), 44–47.

The Philippine Online Chronicle. (2008, 8 October). One Source for Survival: The Benguet Miners' Story. Retrieved from www.thepoc.net/thepoc-features/politiko/politiko-features/469-one-source-for-survival-the-benguet-miners-story.html

PSEC (Philippine Security and Exchange Commission). (2016, 2 November). Press Release on the Reported Mine Tailings Leak in Antamok Mine of Benguet Corporation (SEC Form 17-C). Retrieved from http://edge.pse.com.ph/openDiscViewer.do?edge_no=518ec103e89b0ab972785c150f4cd6d2#sthash.1ArMpy72.dpbs

Philippine Statistical Authority. (2015). Total Population by City, Municipality, and Barangay of CAR as of August 1, 2015. Retrieved from https://psa.gov.ph/content/population-cordillera-administrative-region-based-2015-census-population

Presidential Decree No. 1214. (1977, 14 October). Requiring All Locators Under the Act of Congress of July 1, 1902, as Amended, to Secure Mining Lease Contracts Under the Provisions of Presidential Decree No. 463 Involving Their Patentable Mining Claims. Retrieved from www.gov.ph/1977/10/14/presidential-decree-no-1214-s-1977/

Quirante, A. M. (1624). Expedition to the mines of the igorrotes. In E. H. Blair & J. A. Robertson (Eds.), *The Philippine Islands, 1493–1803* (Vol. 51). Cleveland: The Arthur H. Clarke Company.

Ramos, P. P. (1988). *American Business in the Philippines: The Sociocultural Impact of Benguet Corporation*. (Doctor of Philosophy in American Studies). University of Hawaii, Honolulu.

Rempel, W. C. (1993). *Delusions of a Dictator: The Mind of Marcos as Revealed in His Secret Diaries*. Toronto: Little Brown & Company Ltd.

Republic Act No. 7076 (1991, 30 May). An Act Creating a People's Small-Scale Mining Program and for Other Purposes. Retrieved from policy.denr.gov.ph/basicpol/mincode/ra7076.pdf

Republic Act No. 7160. (1991, 10 October). An Act Providing for a Local Government Code of 1991. Retrieved from http://www.lawphil.net/statutes/repacts/ra1991/ra_7160_1991.html.

The Saturday Evening Post. (1937, 27 March). Gold of Benguet, *209*(39), 26.

Seagrave, S. (1988). *The Marcos Dynasty*. London: Macmillan.

6 Change and continuity in Acupan-Balatoc

> My parents migrated here when they were in their late teens because my father was recruited as a miner. I was born and grew up in Acupan. Each one knew everyone in this area. When the company shut down, thousands of people left. We were among those who stayed because Acupan is our home. The company reopened sometime in 2000, and people started coming back, but it has never been the same because throngs of strangers also entered. We have all sorts of people here now. The newcomers do not treat this place as their own. They leave, new others come in, so we see more new faces. They come here only to make money from mining, with no interest in making Acupan or Balatoc or Virac peaceful and orderly. This not surprising because they are not of this place and never knew this place. It is not their home. It never was.[1]
>
> Mario Balitang (pseudonym)
> A local official of Barangay Virac

This chapter describes the features of community life that is intimately tied to the operations of a mining company. The focus is on present-day Acupan and Balatoc, two places which were former minesite camps that Benguet Corporation (BC) built and directly administered for several decades, but which have transformed significantly following the closure of the Acupan mine in 1992, then its reopening after almost a decade. Acupan and Balatoc adjoin each other and the factors that influenced the growth, development, and transformation of both settlements are the same. Thus, the huge community is referred to as 'Acupan-Balatoc' except in cases which warrant reference to one or the other singly.

Acupan: the old mine and the former minesite camp

The Acupan mine was one of several major underground primary gold mines that formed part of BC's Benguet Gold Operations (BGO). Since 1927, the Acupan mine operated almost without interruption except during the Second

World War and a few years following the war as the mine underwent reconstruction. The mine ceased operating in 1992 due to the decline of gold prices and the flooding of the lower underground work areas caused by the 1990 earthquake (BC, 2002a). At present, the reopened mine, now named Acupan Contract Mining Project (ACMP), has its entry portal and management office in Balatoc. The mine claims were staked along the Batuang Creek but lode mining was in upstream Acupan. Three kilometres downstream is Balatoc, the location of the company's central processing plant into which ores from both the Antamok and Acupan mines were brought through underground haulage tunnels. This centralised mill was installed in 1946. Prior to 1946, each mining area (Antamok, Acupan, Cal Horr, and Kelly) had a separate mill. Balatoc has a relatively larger area of flat land in the narrow valley. Thus, it was also the location of the administration office, warehouse, retail shops, hospital, and main residential area.

As early as 1916, workers and their families experienced the paternalistic management approach of the company president, John Haussermann. The company provided free hospitalisation and medical services, education for workers' children, 'free housing, water, light and fuel and a daily ration of one litre of rice'. The minesite camp included athletic grounds, recreation halls, and a store where camp residents bought clothing, food and other items (De Jesus, 1978, p. 45). Based on a survey[2] in 1985 carried out by the Area Research Training Center of the University of the Philippines-Baguio (Casambre, Pales, Tapang, & Paulino, 1987), BC's housing facilities for its camps[3] in Antamok, Acupan, and Balatoc include the following: 104 bunkhouses (equivalent to 2700 rooms), 103 cottages for rank-and-file employees, 21 cottages, 35 apartment rooms and junior staff bachelors' quarters, 26 cottages, four duplex houses and four apartments for senior staff, and a dormitory and teachers' quarters.

The type of housing given to employees depend on their ranks in the organisational ladder. The senior staff such as managers and executives occupy houses with multiple rooms and fenced backyards. The unmarried junior staff reside in separate dormitory-type buildings. The junior staff with families stayed in single detached or duplex units. Bunkhouses (see Image 6.1) are for the rank-and-file employees, comprising about 96 percent[4] of the workforce, and their families. A bunkhouse is a two-storey building consisting of more than two dozen occupancy units. A unit, intended for two families, is divided by a curtain or other partitioning that the occupants may choose to install. Common toilet and bathrooms for all families in a bunkhouse are in separate buildings. Housekeeping and maintenance roles are distributed to bunkhouse occupants based on a schedule which BC's Camp Administration Department provides.

Image 6.1 One of the remaining bunkhouses in Balatoc, February 2017

Source: Author's research file

With the shortage of company-built housing,[5] many workers constructed their own shacks in areas designated by the company. For this purpose, they procured galvanised iron sheets and other materials from Baguio with their own money. The company, while assisting families by providing electricity and water services, had strictly prohibited workers from purchasing additional building materials to prevent possibilities of enlarging their dwelling structures due to limited space within company-owned and -controlled land. At the security gate on entering Balatoc, company security guards were conscientious in confiscating bags of cement or other items associated with house construction that are without pertinent documents and company permits.

Population data in the mining camps during the pre-war period are not available but Finin (2005, p. 104) collated available information from local newspapers: a) there were more than 17,000 employees in 1937 in five mines in Benguet, b) each mine employed 'between 1,500 and 3,500 workers', and c) among the five mines, Acupan-Balatoc and Antamok employed the largest numbers. There is generally little flat land in Balatoc-Acupan, as in most of Itogon, hence houses had been built on mountain slopes with the majority of them closer to the road. Dense clusters of houses are found at certain spots and intersections, mostly in places where land is spatially larger. After the Second World War, compact clusters of shacks gradually swathed the ramps of the Acupan-Balatoc valley. The clusters grew particularly during the 1970s and 1980s. Draping the slopes towards upstream of the roving canyon, these have been named as Camp Six, Camp Five, Camp 11, Camp 1000, Camp Three, Camp Two, and Camp Lolita. The farthest cluster is Tangtangadan, so named because the settlements is atop the hill making one 'look up' (*tangad*). Downstream in Balatoc towards Virac proper, the clusters of shacks are called Camp Suwerte; Camangga-an, which derived its name from the mango trees (*mangga*) that used to be found in many backyards; Kiangan Village, whose residents are mostly from Ifugao; Bontoc Village, whose residents are mostly from Mountain Province; and Bicol Village, whose many residents came from the Bicol Region southeast of Manila. Located along the slope not far from the administration offices is called Keymen's Hill which used to be the location of cottages for BC's top managers (i.e., key men). Among families who reside in these sub-camps, choosing not to stay in company bunkhouses was not necessarily due to inadequacy of company-provided shelters but also the opportunity for gardening. Others wanted to raise pigs for home consumption and as a source for supplementary income.

Like other minesite communities, Acupan and Balatoc are melting pots where ethnolinguistic affiliation was initially a defining criterion for social interactions. The labels 'highlander' or 'Igorot', 'lowlander' or its sub-groups 'Ilocano', 'Pangasinense', and 'Bisaya' become part of the consciousness of workers, spouses, and children.'[6] The mine camps are also witnesses to how people of diverse linguistic and provincial backgrounds adeptly build affinity with each other. Decades of living together in a bounded spatial area would result to tolerance and gradual blurring of ethnolinguistic divisions towards self-ascribing shared appellations such as *Taga-Acupan* ('of/from Acupan') and *Taga-Balatoc* ('of/ from Balatoc').

A huge camp required an entire company department administering activities for purposes of ensuring peace and order and enabling residents to work and live reasonably well. Camp administration included implementing cooperative activities among residents for cleanliness and sanitation, providing recreation, holding mass celebrations during certain national holidays, and launching sports competition, among others. The company had encouraged and supported income-earning ventures and cottage industries for housewives which included doll-making, basketry, embroidery, weaving, papier-mâché, and sewing. Beginning in the 1930s until the 1990s, there have been at least three to four generations of mine workers and their families who have known Acupan-Balatoc as home. As some workers reached retirement, many of them decided to stay in Acupan-Balatoc or Batuang and other sitios of Virac. This has been convenient to many workers' families because their children have grown up, and quite a few of them had taken their turn of being employed with BC. Other retirees use their severance payments and pensions to invest in improving their shack in Acupan-Balatoc or constructing a house in Baguio. Retirement is a goal of many mine workers. However, the sudden closure of mine operations in the late 1990s caused many workers to lose their jobs before reaching retirement age.

Land in Itogon

BC controls a vast area of mineral lands in Itogon. The Acupan mine in particular controls 223 mineral claims representing a total mine tenement area of 15 square kilometres, which consists of mining claims acquired under several generations of laws, i.e., the Commonwealth Act No. 137,[7] Presidential Decree No. 1214,[8] and Presidential Decree No. 463 (BC, 2002a). Many of the company's claims are patented, acquired under the Philippine Bill of 1902[9] during the American regime, and for which the company has absolute ownership of both surface land and underground.

Land use and land classification have been recurrent issues to residents and local government officials in Itogon. Itogon Municipality has extremely limited 'alienable and disposable' lands. This is due not only to the existence of large corporate mining concessions but also the inappropriate classification of land uses. Layering on 'mineral lands' are 'forest reserves'. An extensive area of Itogon forms part of the Upper Agno River Basin Resource Reserve covered by Proclamation No. 268 that was issued in April 2000. In addition, 88 percent of Itogon's land area also forms part of the 395-square-kilometre Lower Agno Watershed under Proclamation No. 2320 for the San Roque Multi-Purpose Dam. Under the current classification, most of this watershed area is subject to land-use restrictions and prohibitions on land-clearing, farming, animal grazing, gathering of forest products, small-scale mining, and residential purposes. In total, more than 80 percent of land in Acupan-Balatoc is classified as mineral and watershed areas (Barangay Council of Virac, 2005). This means that a vast area of Itogon remains non-residential lands even if generations of people have been inhabiting these lands until the present.

Itogon municipal officials had asked the Mines and Geosciences Bureau (MGB) to determine which mineral lands in Itogon – whether company owned or leased from other claim owners – are already mined out, inactive mining claims, or not economically feasible to mine. Itogon residents were hoping that from such move, a re-classification of lands would proceed and in effect, the mineral lands would be reduced in the municipality's land use map to enable the enlargement of other land uses, particularly residential.

The Department of Environment and Natural Resources (DENR) had finally considered the need for re-surveying and re-classification of lands in Itogon. In 2000, the Itogon Integrated Watershed Management Project conducted a delineation survey to identify built-up areas within mineral claims and the respective occupants of such areas. The results of the survey revealed that there are many residential areas within valid and existing mining claims of various mining companies (MGB-CAR, 2004). These claims had been issued under previous mining laws, the oldest being the Philippine Bill of 1902. In mid-2004, however, the DENR declared that mining firms, which may have shut down their operations, 'still have proprietary rights over the lands that they mined and may use these in new business ventures' (BC, 2003a). The Regional Director of the MGB-Cordillera Administrative Region (CAR) explains that the mining law 'is silent' on the demand of communities to acquire back lands that have been patented. However, the director added, the law requires mining companies to abide by the host municipality's land use plans.[10]

Closure of the Acupan mine

In 1989, the Acupan mine had the first of a series of retrenchment. BC had its huge manpower reduction by over 2,000 of the 6,000-strong BGO work-force. This was followed by three more rounds of retrenchment at the Acupan and Balatoc operations in late 1992 and the first quarter of 1993. The Acupan mine ceased operation in 1992 which necessarily led to the shutdown of the mill in Balatoc. The closures brought down the number of employees to a few scores of company personnel that made up a caretaker management staff who had since occupied the administrative offices in Balatoc. The ceasing of operations was a result of a confluence of events: the decline of metal prices, particularly gold; the 1990 earthquake that caused flooding of underground workings; and the company's inimical conflicts with the workers' union. Furthermore, community protests as described in Chapter 5 against the company's open pit operations in Antamok had not relented.

A mine worker and his family lose entitlement to reside in a mine camp as soon as his employment with the company ceases. Thus, when occupants of bunkhouses and other company buildings left, those who live in self-built shacks have continued to stay in Acupan-Balatoc notwithstanding the fact that the land where their shelters stand is part of BC's patented claims. Following BC's sudden closure, the company sold all materials that could be recovered from buildings and machinery and other assets. Company bunkhouses and cottages were vacated and many were demolished. At present, what remains of company accommodation, based on information from BC, is 75 percent[11] of the former housing facilities before closure. The few bunkhouses now serve as quarters for a small number of BC personnel.

Other units have been rented out for small-scale miners working for the company's ACMP.

Some residents recall that Acupan-Balatoc turned into a virtual 'ghost place' during the early 1990s following the closure of operations. Thousands of people left, making the entire place almost devoid of human life. Lighting along the streets was extinguished. Thefts were widespread even in broad daylight. Major business establishments, public transport, and the company hospital ceased services. The situation started improving in early 2000 when the ACMP began operating. Small retail businesses opened and public transport resumed.

The Benguet Mines Tourism Village

Five years after the Acupan mine was closed, BC converted its mining facilities in Balatoc into a tourism venture – the Benguet Mines Tourism Village (BMTV). Its services include a tour to the Vegas tunnel, a 1942 mined-out underground working area in Balatoc, and a visit to the BC museum that showcases rock samples of gold and ancient gadgets used by traditional miners. During the earlier years of the project, an added feature of the BMTV was the Balatoc Lake, one of the company's former tailings dams, for fishing and cruising activities but it has since closed. The few company facilities that have not been demolished – such as the Villaluna Hall, some cottages, dormitory, swimming pool, and billiard hall – are advertised for vacation and conference purposes. The BMTV has been a modest but significant source of revenues for BC. In mid-2016, business operations were suspended to refurbish some facilities.

As early as 2001, BC's Annual Report indicated the company's aim to develop Balatoc as a new satellite town to Baguio City (BC, 2002b, p. 2). The BMTV illustrates that an area formerly used in mining operation is convertible to another land use. However, the mining company does not necessarily leave.

Pocket mining in Acupan

Throughout the decades of BC's operation, pocket mining within BC-controlled lands was strictly prohibited. Employment with the company is obviously a full-time occupation. It was during the early 1980s when some people began digging in a few hidden sites along hill slopes in Tangtangadan, which happens to be far from the surveillance of company security guards. The few men who started digging mine holes were former BC employees who were either discharged or suspended from work due to being caught committing the infraction 'high-grading'. A person who is charged by the company with the offense 'high-grading' is one who, while at work, concealed and stole ore or tailings that contain an amount or size small enough to hide but with high concentration (thus 'high-grade') of gold. Mine workers' accounts about high-grading are replete with descriptions of various tactics and ploys of concealing gold – in the form of powder (dry tailings) or tiny raw ore (stone) – in every possible orifice in the miner's body. According to De Jesus (1978, p. 75), high-grading is a perennial problem that BC faced costing the company in the order of PhP1 million a month.

By 1983 towards 1984, pocket mining became an activity in Tangtangadan spreading to the nearby Camp Three as the price of gold in the international market surged. More mine holes opened as scores of pocket miners joined the clandestine venture. By 1986 towards early 1987, the number of pocket miners soared in their thousands. Mine holes opened in other sitios such as Camp Lolita, Camp 1,000, Camp Six and Camp Five. During the late 1980s, pocket mining became the main occupation of countless people – men digging, and women sluicing or panning along the river. Thus, it could no longer be a secret activity as BC had little resources to secure its lands and prevent people from digging or panning.

At present, Acupan has numerous work sites of pocket miners whose activities have not slowed down, but rather intensified, especially when BC's large-scale operations ceased. Available data from the barangay *kapitan* (local chief executive) indicate that the population of Acupan-Balatoc collectively in 2003 stood at 4,500. This is an immensely reduced number from about 15,000, representing 2,780 households, in 1990. Following the series of retrenchment of over 5,000 employees towards closure in 1992, there was mass departure of people. However, there were equally large numbers who did not join the exodus, but found ways to stay in Acupan-Balatoc. With no alternative employment, many former BC employees joined the several hundreds of people who have been conducting pocket mining in diverse sites. The number of pocket miners had more than doubled from as little as a few dozens in the 1980s to about 2,000 in the mid-1990s because many people from elsewhere arrived to join the frenzy of finding gold in Acupan.

'Illegal' status of pocket miners

Considering that vast tracts of land in Itogon belong to claim owners, with Benguet Corporation as the largest, most pocket miners are 'illegal' in the sites where they are working. Their status is based on their lack of permits to mine. Consent to mine in an area is required by law for the issuance of mining permits. The matter of obtaining mining permits has been a perennial issue for miners themselves, the MGB, and local government units. The first law pertaining to small-scale mining, Presidential Decree (PD) No. 1899 (1984), aimed to provide the legal basis for issuing licenses and permits. Some miners consider this law 'dead' because they regard it as almost impossible to apply in Itogon which is covered by binding and existing claims of mining companies and individuals, thus there is no way any pocket miner will become 'legal'. Seven years after PD No. 1899 was issued, Republic Act No. 7076 (1991) was passed. It provides for the 'identification, segregation and setting aside of certain lands as 'people's small-scale mining areas'. It also created the Provincial/City Mining Regulatory Board (P/CMRB) as the government's arm to regulate and issue permits to small-scale mining activities. Some pocket miners in Acupan consider the second law to be as inoperable as the first one because Itogon remains covered by mining claims and watershed reservations. The sentiment surrounding the notion of the 'illegality' of mining activities vis-à-vis the intent and practicality of existing laws is expressed by the leader of the federation of small-scale miners' associations who is also the representative of small-scale miners to the Provincial Mining Regulatory Board (PMRB) of Benguet Province:

> Yes, we do not deny that we do not have permits and this makes us illegal. Of course, we would want to become legal, but the root of our problem is the presence of large mining concessions. The requirement for seeking permits is laughable. The law requires that we seek consent of the claim owner. Would you expect Benguet Corporation to say 'sure, go ahead, come and take our ores'? The company has been sitting on its mining claims whether or not it has suspended its operations.[12]

The difficulty of applying the two laws in Itogon, and in broader Benguet Province, also lies in the lack of policy makers' understanding of the traditional character of mining in Benguet which is not the same in gold rush areas such as those in Negros Oriental, Surigao, and Davao Del Norte. To some extent, pocket miners consider their non-compliance with permitting requirements as a levelling mechanism by which they, being the poorer people that they perceive themselves to be, could also partake of the bounties of the earth. The remark of Kilaton, a pocket miner in Camp Five, presents an insight underpinning the 'illegality' of the trade:

> Those who own the big mines may have never known or seen how a mine looks like, but the wealth of this place goes to them. They receive their dividends, don't they? We miners who dig the earth do so to be able to support our families, and sometimes have meat or fish in our pots for our sustenance – just a little quality in our food. Although not every production is gainful, these holes enabled us to improve our living. Why not? Why let only the rich get richer?[13]

The reasoning encapsulated in these statements defies state authority and asserts that wealthy entities must not monopolise minerals. An engineer of BC, expressed the pocket miners' non-compliance to licensing requirements in another light:

> The pocket miners do not really care so much being legalised. Some are not actually bothered whether they are legal or not if they earn money. That is understandable, but the majority of them are totally against the idea of securing permits and being registered because they do not want their output or income to be monitored by the Bureau of Internal Revenue and other government agencies.[14]

In 2012, President Benigno Simeon issued Executive Order (EO) 79 which reiterates the need for small-scale miners to undertake their activities only within a Minahang Bayan. The pocket miners say they will carry on with their activities in Acupan while it is possible and during times when they can. A 34-year-old pocket miner heaves a sigh and says: 'I may not understand fully what the law requires, but I understand enough not to be deceived. The government can not find any area where we could mine. So, we do it right here, right now.'[15]

For many pocket miners in Acupan particularly at present, their need for permit is no longer a real problem because BC has stopped prohibiting their mining activities. A few pocket miners share the view that BC has become more enlightened in accepting that it could not continue protecting company lands indefinitely when there are hundreds of people without work. Others observed that the company had taken the core wealth of Acupan, and is now tolerant towards pocket miners who work on small scattered veins.

The Acupan Contract Mining Project

Following closure in 1992, BC found it imperative to secure its remaining facilities and installations such as the administrative staff offices. The company had been securing its ventilation portals because some pocket miners had been entering the company's underground work areas. Considering that tremendously vast mineralised lands fall within BC's mineral claims, there was too huge an area to protect. However, pocket miners eventually penetrated the company's underground workings. One of BC's managers described the situation:

> There had been diggings in many places. It was becoming impossible for the company to fully secure its property. The company would have

> to spend so much to be able to watch out for the encroachers every day. Our security guards had to chase people who had entered the tunnels. However, there were just too many of them.[16]

Meanwhile, BC's engineering and safety manager explained that the company had no alternative but to legitimise the activities of large numbers of miners:

> The situation was overwhelming. There were countless people needing to find livelihood. At one time, the people who entered the company's underground workings numbered about 300. They had their cooking and sleeping gadgets which meant they intended to remain working underground for a few days until maybe a week when their provisions run out. We in the company then thought that we cannot go on running after them underground indefinitely. We lose. Making the trespassers activity 'legal', which means working with them, was the option left for us. Making them work with us was the way to stop the cat-and-mouse bustle.[17]

The company managers found it feasible to engage in a partnership with pocket miners for two main reasons. First, establishing a sharing venture with pocket miners could be the only way for BC to continue mining because the option of re-hiring employees was not financially viable. Second, forging better relations with the community was necessary and the only way for both parties to co-exist. BC managers understood that there was no way to stop people from carrying out their mining activities, prohibited though they may be.

Conceptualising the ACMP started in 1999. It was to involve about 700–800 small-scale miners who were to carry out production activities through a contract arrangement with the company. The official reopening of the project was held in 2000. Adopting the name Acupan Contract Mining Project, the company intended the project to be a 'low-capital, low-cost venture' (BC, 2002b). Commercial operations began in 2002. The company reopened four levels of its Acupan underground workings: Level 2000, Level 1875, Level 1700, and Level 1500. Level 2000 is the portal in Balatoc while Level 1500 is the portal in Acupan.

The ACMP has been regarded at the outset and publicised both by BC and the Mines and Geosciences Bureau (MGB)-Cordillera Administrative Region (CAR) as 'a model for partnership between a large-scale corporate mine operator and small-scale miners' (BC, 2005, pp. 17, 34; Gomez, 2004). The small-scale miners, to be qualified in providing production services for the company, are required to form cooperatives or associations. Regarded as a paradigm shift from conventional mining operation, the partnership has been anticipated to be mutually beneficial to both the company and the community. On the part of the community, a livelihood source is provided which also makes the pocket miners' activities 'legal'. This means they do not have to sneak into company underground workings. On the part of the company, its remaining gold ores are extracted at a profit with the contract miners serving as the workforce for production.

The partnership involves a third party, the government, represented by mainly the MGB, supported by the Environmental Management Bureau, and the local government, i.e., the Virac Barangay Council and the Itogon Municipal Government. The MGB's main role is to create the mechanics, rules, and regulations that would govern the cooperative arrangement. Its major goal is to address unsafe and destructive mining practices. As conceptualised, the company would provide mine planning and development, engineering, geology, safety and pollution control services.

The mining method employed is the '*camote*' mining which pocket miners are most familiar with, i.e., excavating narrow-vein deposits (likened to digging for camote, the vernacular for sweet potato) to collect ores. The ores are placed in sacks and hauled manually or loaded onto flatcars to the lode shaft for the upper levels. They are eventually brought out to Level 2000 either by manual hoisting or by using the pipeline at the lode shaft. Locomotives haul the sacked ores to the mill for processing. The company spent around PhP15 million for purchasing and installing a 50-TPD (tonnes per day) carbon-in-leach/carbon-in-pulp plant to provide central processing services for both processes of extracting higher grade ore from underground mining operations. By providing a centralised ore processing unit for all small-scale miners, BC claims that it makes possible for the compliance to environmental laws in regard to processing (BC, 2003b).

The ACMP started commercial operations in 2002 with the implementation of the first mining contracts. After three months, the scheme shifted to production sharing: the ore that was mined is shared between the two parties in which 60 percent is allotted to the contract miners and 40 percent to the company. In March 2003, ACMP had its second contract bidding for a two-year mining program, with seven accredited mining cooperatives allotted 14 mining areas distributed among themselves. There has been some reluctance among miners about the 60–40 ratio. Thus, the groups of miners earlier agreed among themselves to boycott the bidding in the hope that with no available bidders, the company might reconsider changing the ratio to something better that would allow the miners to get more than 60 percent. However, some of those who concurred to boycott the bidding participated in it just the same after they felt they understood the sharing scheme. Thus, the ratio has remained unchanged.

The contractor who finances a group's mining activities sees to it that its group meets the production quota. In this scheme, the contractor serves as an intermediary – that of an administrator, coordinator, and financier – between BC and the miners. The company designates production quotas that each contracting group must produce. Anything beyond the quota is something the group members could keep for themselves, including ores from high grade gold veins which they may be fortunate to hit. The contract miners do not talk about the bonanzas that they could be lucky to chance upon. As the kapitan describes, working for ACMP is both winning and losing:

> The contract miners are sometimes the losers but sometimes they are the winners. They have bad days and good days. Sometimes they hit a high-grade (ore) and that would be a big win for them. Sometimes it's just continual work and they simply receive their regular share. But in the end, they do not get zero.[18]

A few sources made comparison of the previous situation when BC had employees, and the current setting in which ore production is contracted out by the company. A retired employee who had worked for BC from 1968–1993 observed:

> The old system in which the company had employees was better. This time with the new setup of having the ACMP, the company is freed of obligations and accountabilities including formal indemnification for miners who meet accidents or, God forbid, get killed. On the part of the contract miners, they get no entitlement for vacation or sick leaves. But we are maimed by the fact that the ACMP offers employment to people.[19]

Given the perceived weaknesses in the contract mining arrangement, the ACMP has somehow provided peace and order in the community. BC management endeavours to improve its system. The former kapitan of Virac notes that the ACMP has helped employ formerly jobless people who otherwise might be simply remaining depressed, engaging in drinking liquor, and causing trouble. The current kapitan observes that in the last five years, some contractors have been recruiting people from their own provinces, which is contrary to the idea of employing local residents of Virac.[20]

The ACMP versus pocket miners in upper Acupan areas

Not all pocket miners in Acupan work for the ACMP. Hundreds of others have maintained their mining activities, independent from any influence of BC. This is the case since the mid-1980s in the upper areas of Camp Five all the way to Tangtangadan. One of the underground areas which BC opened for the ACMP is Level 1500, which happens to be closest to the upper areas of Acupan. In August 2002, a number of residents in the area set up a barricade to prevent contract miners of ACMP from working at Level 1500. The opposition by residents of Camp Two, Camp Three, Camp 11, Camp 1000, and Camp Lolita to ACMP's mining activity in the site was based on their belief that the underground workings would collapse and the water sources in the surrounding communities would dry up.

The barricade was led by officers and leaders of the Acupan Upper Camps Community Livelihood Association, Inc. (ACUCCLA) whose members were themselves former miners and supervisors of BC. They represent residents of the Acupan upper areas including Abucayan which is part of Barangay Ampucao. In response to the barricade, Itogon municipal officials instigated a dialogue between representatives of BC and the ACUCCLA. The Itogon municipal council approved Resolution No. 306–02 which called for an extensive and comprehensive technical study showing the available ore reserves and excluding areas identified as 'no mining zone' including Level 1500. The Provincial Board of Benguet, in support of the action by the municipality of Itogon, expressed its official stand that it favours the closure of Level 1500 because of the potential danger unless a comprehensive study is conducted (Sunstar Baguio, 2003). The barricade lasted for several months until the end of 2002 when ACUCCLA leaders compelled the MGB to declare part of Level 1500 a Temporary No Mining Zone (TNMZ). Since then, the residents kept watch over the Level 1500 portal to prevent entry of people whom they consider intruders.

The technical staff of MGB-CAR explain that Level 1500 has some areas that are mineable, separate from the designated TNMZ. The company is keen to forge a contract with ACUCCLA or any of the residents to mine a certain site within Level 1500. To the ACUCCLA members however, Level 1500 is no longer feasible for mining and the tunnels have long been abandoned and without electricity. In addition, reinforcements are no longer maintained.

In 2006, BC somehow got a contractor who was willing to mine a designated area within Level 1500. In the company's attempt to maintain partnership with the community, the person who was given the contract to carry out the project was a resident of Camp Five. The designated contractor and his men were about to start mining activities but were stopped by residents of the upper camps in Acupan. In the end, the contractor was not able to proceed with the project because he was ostracised by the residents and dismissed as 'selling out his town mates' and instead favouring the company. ACUCCLA leaders petitioned anew the MGB to restrain BC from continuing to mine the site. The petitioners claimed that the provincial government, municipal government, and Barangay Council of Virac were not consulted before any project activity could commence aside. In addition, the petitioners argued that Level 1500 is not covered by any valid contract, which is regarded necessary before contractors could proceed with mining the site. The MGB turned down the petition, declaring it was devoid of merit. At present, Level 1500 is being mined by BC through ACMP, and has continued until the present.

Current challenges in Acupan-Balatoc

Among the contract miners of the ACMP in Balatoc, the reopening of the Acupan mine in 2002 was a relief to economic doldrums and desperation. Their work as contract miners brought them the much-needed income. Some hope to keep their livelihood for as long as they need it although they are aware the ACMP will not be around forever. With BC's reopening of mining activities, Acupan-Balatoc has become a hub of people that include the indigenous Ibaloy of Virac, old settlers, newcomers, former BC employees, non-employees, sojourners, and some 'wanderers'. Several miners working for the ACMP – particularly those who act as 'lead men' or financiers of the working groups – have recruited their own kinsmen from the provinces. With the increase of people who are considered 'newcomers', many old-time residents in Acupan-Balatoc feel that it is difficult for a strong sense of community to develop. In the description of some Ibaloy and Kankanaey residents, many newcomers do not treat Acupan-Balatoc as their home but merely as a place to earn money, leave, and then come back for another sojourn. The recent waves of migrants have important bearing on the changing dynamics in Acupan-Balatoc as a community. The increase in their numbers is attributed to some mine contractors hiring their own kinsfolk to work for the ACMP. Similarly, a barangay official notes that new workers from the lowland provinces of Zambales and even as far as the Visayas are around because some ACMP contractors brought them.

A considerable number of residents who were born and grew up in Acupan-Balatoc belong to the ethnolinguistic groups *Ifugao*, *I-Bontok* and *I-Aplay* (northern Kankanaey) from Mountain Province. They and their parents were among the larger number of families in Acupan-Balatoc before the Acupan mine's closure. Their numbers evolved two sub-camps called 'Kiangan Village' (mainly Ifugao) and 'Bontoc Village' (mainly Mountain Province). Since 2002, there has been a pronounced increase of *I-Kalinga* as they come to work for an ACMP contractor who hired them. There is evidence that the Ibaloy and Kankanaey feel overtaken. A retail store owner in Acupan, sounded griping when she said to me:

> Your fellow I-Bontok invade Acupan. The old mine closed, and they are still around. They dig the gold here, earn a lot of money, then build

> their big houses in their own hometowns. It is your hometowns and your province that improve, not Itogon, but we people in Benguet do not do what you people do. We do not go to your province to exploit your resources.[21]

Among the pocket miners in upper Acupan who are not with the ACMP, their most pressing concern is the unstable tenure over their mining sites. They claim that their ability to continue mining is partly a result of their strong representation, through the president of the small-scale miners' federation, to the Provincial Mining Regulatory Board (PMRB). Some MGB staff describe the federation president's ability to wield influence, using wit and perspicacity, on the PMRB to be 'soft' on the permitting requirements. At the MGB-CAR, the officer for small-scale mining observes that the issue about livelihood gives no reason for the PMRB to be strict. In their mining ventures, pocket miners in the upper areas of Acupan seriously need water as it is extremely important in running ball mills and sluicing gold. Many pocket mining groups incur huge expenses to reticulate water from rivers or streams which are located outside Acupan. In addition, timber is needed for ground support in their tunnels. Most areas of Itogon has been denuded by pocket miners themselves. Thus, they buy timber from faraway towns of Benguet like Tublay and Atok. Pocket miners are aware they are required to provide containment ponds for their mine tailings. However, a considerable portion of mine wastes still find their way to the gullies, and eventually flow to the local Ambalanga River and ultimately into the Agno River,[22] causing siltation of waterways and the company-built diversion tunnels.

Some Ibaloy and other long-time settlers of Barangay Virac describe new settlers as not interested in becoming part of the community. They are not concerned about how Acupan-Balatoc is wasting away. Since the closure of the mines in 1992, the road from Balatoc to Acupan was very unpleasant to travel, appallingly dusty in the summer, and dreadfully slippery during the rainy season. In 2013, some funds from the local council augmented by some allotment from the provincial government made possible the concreting of the 3-kilometre road from Balatoc to Upper Acupan.

A mother observed that the proliferation of ball mills and leaching pads in Acupan, outside the purlieus of BC's ACMP, create constant noise and pollution. However, she quickly adds: 'residents have grown tolerant with the reality that living in a mining community includes getting used to the lack of neatness of surroundings and constant noise and dust in the place.' The once scenic landscape of Acupan is now a vista of hillocks dotted with adits and makeshift shelters. One pocket miner estimates that in Camp Five alone, there are over 90 mine holes. The other major sitios like Tangtangadan and Camp Three would have hundreds. Other residents quipped about the growing garbage problem. Considering that the Barangay Council of Virac has no existing system of garbage collection in either Acupan or Balatoc, solid waste continues to be thrown into the creeks. Semi-permanent structures for shelter have blotched the hills as more people have come to work and stay in shacks commonly located right where they mine.

The barangay health workers' survey in 2003 (Barangay Council of Virac, 2005) included an open-ended question asking household heads in Barangay Virac about their top three main problems in their family. From the various responses, the primary concerns of households are about: a) money, livelihood, and income source; b) health protection; and c) proliferation of vices. Other significant concerns reported pertain to housing particularly high rent and lack of a permanent house, poor sanitation, and shortage of water supply. While there is no updated information available, barangay officials at present observe that these challenges remain the very same major ones they face.

The local government, especially the barangay council, also bear the unwanted effects of mine closure. In 2005, my interviews with barangay officials gathered expressions of resentment against jobs in the ACMP being taken by outsiders instead of original residents of Acupan and Balatoc. Data from interviews in February 2017 indicate an even stronger antipathy. An added frustration to the barangay council is the situation in which taxes could not be collected from the ACMP miners because the company already pays excise taxes to Itogon municipality. The barangay council is keen to collect some revenue from the hundreds of pocket miners in Acupan-Balatoc. For this, it has recently come up with an ordinance that every sack of ore will incur a 'tax' of PhP0.50, although it will not be easy keeping track of the production of pocket miners. The barangay council has also required each ACMP miner to apply for a working permit.

A mining engineer at MGB, who grew up in Balatoc, expressed apprehensions that with the continual mining activities by both the ACMP contract miners and the huge uncontrolled number of pocket miners, the place might experience an ecological disaster in the years ahead. BC undertakes mine planning for the ACMP and the MGB carries out safety inspections. While this is true, some residents remark that contractors and their men are simply out to procure their quotas. In the same way, the pocket miners in Acupan face the daily challenge of earning as much income as they possibly could with little regard of environmental obligations they need to fulfill.

In 2005, the mayor's sentiment speaks of the weight of challenge faced by the Itogon municipal government:

> For decades, Benguet Corporation has indeed stirred business in this municipality, employed people and built the roads which the company itself used to the maximum, until the time the company closed down.

> The municipality has faced many problems since then. It is not fair that we are the ones being confronted with the lack of jobs for people or the absence of other industries that might be viable in Itogon. There should have been a system in which the company had to address the problems brought about by the closing down of its operations, and not leave the problems to us.[23]

The present kapitan of Virac says the nature of the burden is the same until the present – the need to manage the social changes indirectly brought about by BC, particularly the arrival of many people. He explains that controlling the influx of people will be possible only with the help of BC. He notes that the company must be genuinely strict in screening people who enter Acupan-Balatoc right at the checkpoint gate. He feels that the biggest threat in Barangay Virac is the possibility of Acupan-Balatoc becoming a gold rush town. He adds: 'The company is too lax in their security check. BC is our partner in this barangay so it must help us protect Acupan-Balatoc from uncontrolled entry of new people.'[24]

Notes

1 Author's interview, February 2017, Camangga-an, Virac.
2 I was part of the research team that carried out the intensive population mobility survey of ARTC-UP Baguio, covering areas of mining (Itogon), agriculture (La Trinidad) and urbanisation (Baguio).
3 There is no disaggregated data for each of the camps.
4 Personal communication (via e-mail) with BC chief of staff to the officer-in-charge, 18 April 2017.
5 See Alabanza (1991) for relevant information on housing conditions in Benguet Gold Operations communities.
6 Finin (2005) presents a pithy discussion on the mines as places where 'Igorots discover themselves' through their interactions with Americans and other Filipinos, particularly lowlanders.
7 This law, also known as the Mining Act of 1936, granted leasehold rights to mining claimants.
8 Presidetial Decree (PD) 1214, issued by President Marcos on 14 October 1977, required all locators under the Philippine Bill of 1902 to secure mining lease contracts involving their patentable mining claims as provided by PD 463.
9 The Philippine Bill of 1902, passed by the U.S. Congress, allowed the open and free exploration, occupation, and purchase of minerals and the land where they may be located. Patents and certificates of title were issued in favour of the locator or mining claims owner for patented mineral lands (Cagampang-De Castro, 2009).
10 Author's interview with MGB-CAR Director Neoman de la Cruz, Nov 2004, Baguio.
11 Personal communication (via e-mail) with BC chief of staff to the officer-in-charge, 18 April 2017.

12 Author's interview, October 2005, Baguio.
13 Author's interview, February 2017, Camp Five, Acupan.
14 Author's interview, November 2004, Balatoc.
15 Author's interview, February 2017, Camp Three, Acupan.
16 Author's interview, November 2004, Balatoc.
17 Author's interview, November 2004, Balatoc.
18 Author's interview, March 2005, Balatoc.
19 Author's interview, March 2005, Balatoc.
20 Author's interview with Barangay Kapitan Noel Bilibli, February 2017, Virac.
21 Author's interview, May 2005, Campo Uno, Acupan.
22 The Agno River is a major source of irrigation for the agricultural towns in lowland Pangasinan Province.
23 Author's interview with Mayor Mario Godio, February 2005, Poblacion, Itogon.
24 Author's interview, February 2017. Camanga-an, Virac.

References

Alabanza, M.A.E. (1991). *The Experience of Crowdedness in a Philippine Mining Community*. (ProQuest Dissertations Publishing). Michigan State University.

Barangay Council of Virac. (2005). *Barangay Profile of Virac*. (Unpublished). Virac, Itogon, Benguet.

BC (Benguet Corporation). (2002a). Acupan Gold Mine. Retrieved from www.benguetcorp.com/html/Acupan%207-5-02.pdf

BC (Benguet Corporation). (2002b). *Benguet Corporation 2001 Annual Report*. Makati, Philippines: Benguet Corporation.

BC (Benguet Corporation). (2003a). 2003 First Quarter Report Benguet Corporation and Subsidiaries. Retrieved from www.sec.gov/Archives/edgar/data/11290/000001129003000007/benguet.txt

BC (Benguet Corporation). (2003b). *Benguet Corporation 2002 Annual Report*. Makati, Philippines: Benguet Corporation.

BC (Benguet Corporation). (2005). Benguet Corporation 2004 Annual Report (SEC Form 17-A). Retrieved from www.benguetcorp.com/html/BC%202004%20Annual%20Report.pdf

Cagampang-De Castro, S. M. (2009). Acquiring mineral rights in the Philippines: Updates on mining law and jurisprudence. *Philippine Law Journal*, *83*(3), 609–671.

Casambre, A. L., Pales, D. C., Tapang, B. P., Jr., & Paulino, B. M. (1987). Population mobility and development in the mining sector. In Area Research and Training Center – Cordillera Studies Center (ARTC-CSC) Team. (Ed.), *Population Mobility Survey in the Baguio Area* (pp. 242–393). Baguio City: ARTC-CSC, University of the Philippines College Baguio.

De Jesus, E. C. (1978). *Benguet Consolidated, Inc. 1903–19878: A Brief History*. Makati City, Philippines: Benguet Consolidated, Inc.

Finin, G. (2005). *The Making of the Igorot: Roots of Cordillera Consciousness*. Quezon City: Ateneo de Manila University Press.

Gomez, A. V. (2004, 17 March). *Benguet Corporation's Acupan Contract Mining Project (Mining in Partnership with the Community)*. Paper presented at the International Symposium on the Diversity of Mining and Sustainable Local Development:

A Meeting to Study the Business Practices of Small-Scale Gold Mining in Benguet, Philippines, University of the Philippines, Baguio City.

MGB-CAR (Mines and Geosciences Bureau – Cordillera Administrative Region). (2004). *Paltek di Cordillera, 4*(13). Baguio City: MGB-CAR.

Presidential Decree No. 1899. (1984, 23 January). Establishing Small-Scale Mining as a New Dimension in Mineral Development. Retrieved from www.gov.ph/1984/01/23/presidential-decree-no-1899-s-1984/

Republic Act No. 7076. (1991, 30 May). An Act Creating a People's Small-Scale Mining Program and for Other Purposes. Retrieved from policy.denr.gov.ph/basicpol/mincode/ra7076.pdf

Sunstar Baguio. (2003, 9 March). Benguet PB Bats for Closure of Mine Tunnel. Retrieved from www.sunstar.com.ph/static/bag/2003/03/09/news/benguet.pb.bats.for.closure.of.mine.tunnel.html

7 Drawing lessons from Itogon and Tuba for mine closure challenges

> If I were to start a new mine, I want to start from a clean slate. I will buy the land, no matter how much it costs. That way, the land is my own property, and I will have no problem about people claiming surface rights. It will be neat, and the gains will be long-term. As a mining engineer, dealing with rocks and production has always been fun. What is exhausting has been the need to deal with claims of surface owners who themselves fight each other because their claims are overlapping. It is in these conflict situations where politicians come in, and more complications set in. We pay all legitimate claims for compensation, following the principle 'Pay – give, and give all.' But they themselves must settle their bitter quarrels first.
>
> Emiliano Andiso[1] (a pseudonym)
> A Mine Operations Manager

The accounts of Benguet Corporation (BC) in Itogon, and Philex Mining Corporation (PMC) in Tuba, showcase three different mines with varying experiences of mine closure. The Antamok open pit mine closed in 1998, and the company has been attending to rehabilitation requirements for its open pit and its vicinities. The Acupan underground mine closed in 1992 but reopened through contract mining, partly in response to care-and-maintenance concerns, and resumed commercial operations in 2002. The Padcal mine is projected to close in 2022 and the company began a mine transition plan that is being updated. All three cases demonstrate that the mine closure process is an intense locus for competition, opposition, arbitration, and compromises among various actors who assert their respective claims and interests. The interactions sometimes reach mutual points, but actors also collide due to incompatible values and objectives, power imbalance, and differing lifeworlds across cultural, political, and spatial domains. Overall, the communities' participation in the mine closure process is not necessarily in relation to charting the 'post-mining future', but rather, in relation to their immediate concerns – the here and now. The experiences of BC and

PMC offer key messages for understanding the complex socio-cultural, economic, political, and business realities that make up the social terrains of mine closure, and therefore configure the nature and tempo of closure strategies and related activities.

The broad national domain of the social terrain

The Philippines has had a long history of mining. The timing, pace, and structural framework of minerals exploitation in the country have been shaped by colonial laws which have been honed by subsequent administrations whose economic policies were all for inviting more foreign capital into the country. The neo-liberal economic paradigm underpins the provisions of the country's current mining policy, the Mining Act of 1995. It is a manifestation of a globalised market and a consequence of the neo-liberal policy governance that has saturated most developing countries beginning in the 1990s. Similar to Chile and Argentina (Haslam, 2007), as well as Indonesia (Bhasin & McKay, 2002), the Philippine government forges financial and technical alliances with transnational firms to achieve broad economic goals. The autonomy of government policy in mineral development is restrained by the demands of globalisation, 'the growing liberalization of international trade and investment, and the resulting increase in the integration of national economies' (Griswold, 2000). This reality means that transnational mining companies will continue to be invited and enter the Philippines, and the minerals industry will remain export-oriented. Thus, the dynamic but often acrimonious interlocution and positioning between the minerals industry and mining oppositionists will continue as the latter advance their criticisms of the industry and support bills that have been filed at both Senate and Congress which aim to supplant the Mining Act of 1995.

The enduring antagonism to mining is a key feature of the general social terrain of mine closure in the Philippines. The numerous mining operations in previous decades that had abandoned their sites without rehabilitation have become the defining example of mining in the country. This has created the plain and persistent, but unfortunately exiguous understanding that new mining projects will not be different from those of the past. This view, which is widespread, has gained credence from the occurrence of tailings disasters and other mining-related misfortunes in recent years.

Long before the Mining Act of 1995 was enacted, the country has had the necessary policies requiring rehabilitation of mining areas (PD No. 1198, 1977) and fees on mine wastes and tailings to compensate for damages (PD No. 1251, 1977). Despite these laws, corporate cultures in the minerals industry have not built real commitment to rehabilitating former mine areas because, during the Marcos regime, enforcement of regulations 'were at best lax' and political interventions with cronyism were at play while communities were largely indifferent (Domingo, 2003, p. 4).

The Mining Act is supposed to help correct the blunders of the past. Following the resolution of the Act's constitutionality 10 years after its enactment, the key focus of the Department of Environment and Natural Resources (DENR) and the minerals industry has been towards improving governance mechanisms related to attracting the much-sought after foreign exchange. The jubilation of both the government and the minerals industry about foreign capital flowing into the country, resulting from the mining revitalization program, has not been fully matched by the required cadence in correcting weaknesses in environmental governance. Thus, the haunting ghosts of the 'sins of the past' have persisted, compounded by perennial questions about equity in benefits from mining, among others. Needed actions have started to be implemented which include Social Development Management Programs, and Environmental Work Programs of operating companies. A notable development is the Philippines becoming an Extractive Industries Transparency Initiative (EITI) candidate country in 2013. During 2015–2016, many large-scale mines complied with the DENR directive to obtain ISO 14001 Certifications particularly for environmental management system. However, it remains the case that these initiatives are reactive, which means that companies were compelled to improve systems to obviate damage. Not surprisingly, anti-mining animosity – based on environmental and nationalist grounds – that has deeply pervaded civil society groups has already gained traction, and has relentlessly posed onerous challenges to large-scale mining projects.

In their study on institutional arrangements to build synergies between mining, livelihoods, and development in Peru, authors Bebbington, Bury, & Turner (2009, p. 17296) found that 'serious sustainability problems will emerge' when mining expansion precedes institutional innovation. In the case of the Philippines, the required institutional innovations to improve environmental governance are certainly long overdue. Bebbington et. al. (2009) suggest that social conflict, if substantially mediated, offsets power imbalance and brings about sustainability-enhancing institutions. The case of the Philippines is something to behold as we wait and see. The opposition to mining is not solely about the misdeeds in previous decades that brought about negative mining legacies. More fundamentally, the obstinate questions that the minerals industry faces are about exploitation of the country's minerals by whom and for whom. Other issues include more transparency and accountability needed in the industry, and sharing of revenues between the national and local governments (Ph-EITI, 2016), among others. Meanwhile, few people within Itogon and Tuba would have thought about these issues to bear on their circumstances. As the foregoing discussion presents, the communities have their own immediate, specific, and locality-based concerns.

Mine completion and alternative land use

Mine closure planning, decommissioning, and rehabilitation are ideas that have become part of the vocabulary of the DENR and key officers within mining companies as well as civil society and community representatives involved in mine monitoring and mine rehabilitation. However, decommissioning and rehabilitation towards the ultimate relinquishment of a former mine area to another user remain to be imperceptible to many people. They are notions that are unfamiliar, if not abstract or unreal, particularly in places where new or future mining projects may be planned. The plain idea of 'mining as a temporary land use' remains hardly proven, just as the concept of handing over a former mine land to another user has yet to be tangibly demonstrated.

The notion of planning for closure – at the time a mine commences – is little appreciated by local governments and communities, and others when they are far more occupied with the immediate than the imagined. They have less time to bother with a future that has not happened, but more with concerns that are real and at hand: acquiring benefits from mining at present, or rejecting a new or potential mining project on environmental, social, economic, or other grounds.

Ideally, as stated in Chapter 1, mine completion is the endpoint of mine closure. A mine is considered complete when the former mine land/area is finally passed on to the next land user who would then assume the responsibility over the land; relinquishment is the consummation of closure activities (AGDIIS, 2016; Peck et al., 2005). In the Philippines, the situation of land use and classification is fraught with complexities. For example, vast lands in Itogon have ossified as mineral lands and watershed reservations regardless of the decades-long occupancy of thousands of residents on those lands. Considering the huge populations in areas where mines have operated, it is highly probable that a former mine land bequeathed to another entity (e.g., a government agency or a local government), which has never happened to date, will become 'residential' in use, regardless of classification.

Another challenge concerning land uses is the nature of mining and its products. In the Philippines, some owners of the unrehabilitated mines have considered reopening the sites. After decades, however, the plans have not materialised, and the mines have become liabilities. The nature of minerals as products impinges on definitions of temporariness or permanency of closure. What may be considered a high-grade ore one time may be regarded differently at another time. As mining, milling and smelting technologies continue to develop, and metal prices soar and dive, the dividing line between ore and muck, in any given location, invariably changes based on the ever-volatile metal prices in the international market, and the costs of reopening former working areas. Thus, decisions about converting a mine land to another use, in the context of mine completion, is not always relevant. The reopening of the Acupan mine is an example that a closed mine can reopen.

Progressive rehabilitation of mined-out sites is being carried out by some operating mines at present, although the reclaimed areas are 'still subject to the mandatory relinquishment by contractors provided by law' (MGB, 2017). Nonetheless, there remains no case in the Philippines where a mining company has fully rehabilitated an entire former mine area, handed it over to the next user, then walked away.[2] In the case of PMC, the idea of relinquishing the mine area and associated facilities to local governments (barangays of Ampucao and Camp Three, Tuba municipality and Itogon municipality) is a goal. Officials in local governments themselves are not sure how far this goal may be as they themselves wonder about resources and capabilities required to manage the area and facilities that will be bequeathed to them. While PMC has set schedules for a series of activities in rehabilitating and decommissioning specific sub-areas and component structures (e.g., residential areas, tailings ponds, company offices, mining and mill complexes, etc.), coordinating with local officials at barangay and municipal levels has also carried on. However, elected officials who have defined terms of office are more preoccupied with the pressing concerns such as the shares of tax revenues from PMC that need to be divided between two municipalities.

Regarding BC's Antamok mine, relinquishment depends on business expediency and other factors. Vast areas of land from which BC's mining production had taken place are the company's patented claims, i.e., private properties. BC maintains every chance of resuming work production in the Antamok mine depending on the prospects of metal prices. Based on 1999 company data, the Antamok area has an estimated resource of about 12.4 million tonnes averaging 3.45 grams of gold per tonne (BC, 2003, p. 3). Part of the company's patented claims include Surong and Luneta where ore reserves 'had remained untouched since the mine's closure' (BC, 2010, p. 13), but the two densely populated settlements are well-entrenched communities. BC faces moral considerations prevailing over commercial goals.

Considering that the open pit has not been backfilled since operations closed almost 20 year ago, BC has dealt with the constant need to prevent numerous pocket miners from encroaching into the pit and the vicinities. Thus, the company has submitted to MGB-CAR its proposal to set aside an area where pocket miners will undertake their activities. The company is voluntarily giving up 0.52 square kilometres of its patented claims for a Minahang Bayan in Itogon. This will be a partial form of 'relinquishment'. In Acupan, BC has been able to develop an alternative use of some components of its huge properties. It developed a 1942 mined-out underground working area, together with some company facilities, as part of an ecotourism venture. There was no handing over of the tunnel and facilities to another user. Rather, BC itself has run the venture as part of its own business. This was possible because the company has diversified into non-mining businesses such as real estate development. On the other hand, other patented mineral lands of BC in Acupan have remained as mining sites of many groups and individuals whose illegal status remains unresolved. The company could only do so much to stop pocket miners from encroaching. This is toleration of other users of mineral land, not relinquishment.

Communities, relationships, and human agency

Communities vary in the form and extent of their relationship with a mining company, local governments, NGOs, and others. One factor which effectively shapes the nature of mine closure mechanisms, but is hardly discussed in the literature and therefore scarcely understood, is the political-administrative aspects of governance. A community is a constituent sub-unit within a network of power relations and decision-making processes. Within the formal governance structure which involves defined political jurisdictions, a community can be a sitio (e.g., Antamok) or a group of sitios (e.g., those surrounding PMC mine area), a barangay, a minesite camp that strides between two barangays (e.g., Philex), or two contiguous minesite camps within one barangay (e.g., Acupan and Balatoc). The formal delineation of political-administrative domains has major implications for implementing policy and programs relevant to closure planning and implementation.

Recognising and understanding political-territorial jurisdictions help mine closure planners demarcate the scope of action and identify key actors (e.g., local politicians) to work with. Considering that communities are the lowest sub-units within the administrative setup of governance, they can take unnecessarily circuitous paths to obtain the attention of those who control the mechanisms of decision-making regarding allocation of goods. In many instances, the mining company is regarded by townsfolk or villagers as a channel for, ultimately, resource distribution in their favour. In other instances, groups turn to NGOs to articulate their interests, and may subsequently realise their goal.

Antamok and Acupan-Balatoc

History and tradition have shaped the relationship of communities with the mines. Antamok and Acupan-Balatoc had been created by histories of labour recruitment and mineral production. As productive activities had to commence and carry on for several decades, BC modified the physical, economic and social landscapes of Antamok, Acupan-Balatoc, and the broader Itogon municipality. BC has established the predominant land use of vast lands in the area, notwithstanding the non-secure occupant status of many residents until the present. After closure, the social landscapes of both Antamok and Acupan-Balatoc have significantly changed, with BC still an influencing and necessary actor in matters concerning mining as an income source in Itogon.

The company is in an advantageous position of having complete control over what it decides to do with its facilities because they are within its patented claims. Such a secure type of tenure over surface land and minerals underneath is a result of colonial laws that were purposely legislated to seal off stakes and claims of capitalists during the American period. BC is able to pursue on its own terms how lands under its ownership and control is to be used. Government policy is inadequate in dealing with situations where strategies for 'rehabilitating' a mine area is contingent mainly on what the company resolves to do or not do. As pointed out earlier, the principle 'mining is a temporary land use' finds inadequate meaning when the land is privately owned by a mining company.

The changes and continuity in Acupan-Balatoc also demonstrate that even with the economic and political leverage of BC to steer the course of events according to its corporate objectives, the people in Acupan-Balatoc, are not bereft of options. Pocket miners in the Upper Acupan areas, through individual and group initiatives, continue operating 'illegally' within company-owned and -controlled lands. Similarly, the opening of the ACMP is a product of BC's encounter with another group of miners who also defy the company's legal ownership of mineral lands. BC had no recourse but to work with this group of miners by developing a scheme to make profits from company lands being encroached upon and minerals getting stolen. This gave birth to the ACMP.

Padcal and surrounding sitios

PMC's occupational community, Philex, has symbiotically developed with the company's decades-long operations. The dislocation of mine workers and their families is not only economic, but also cultural and emotional which cannot be fully gauged. PMC's attempt to alleviate the dislocation is through generous severance payments. The more than 1,200 remaining workers face varying prospects in the subsequent years as PMC continues to trim down the workforce. Meanwhile, the sitios around PMC's mine and its occupational community have endeavoured to acquire 'gains' from the company in the light of the mine's eventual cessation of operations. The concerns about loss of land and water, which are interlaced with claims to locality and indigeneity, may have been relatively quiescent during the first 40–50 years of PMC's operations. However, they re-emerged as imperative and urgent demands beginning the time PMC announced the mine's impending closure. Being the people currently occupying Padcal, residents of Philex and surrounding sitios have a valuable opportunity to influence the post-mine scenario. The extent of their influence on the future scheme of things depends on how they take advantage of this juncture.

Benefits from mining

People in mining-affected communities are aware that they too should partake in the benefits from mining. They have ideas of what is fair or not fair. The residents in sub-villages surrounding Philex perceive that they receive mere trifles while the mining company amasses profits. The locals regard the migrant mine employees as receiving far more benefits from the company. Mine closure strategies should incorporate schemes of addressing equity issues, sometimes across different sectors, considering that the very process of mine closure itself becomes a locus for disagreements and divisions.

A mining operation creates expectations. The first generation mining companies unwittingly became the provider of basic industrial and social infrastructure, such as roads, power and water services, schools, and hospitals for communities. These expectations are justified. A mining company built communities not for the sake of doing it but because the goal of business viability dictates doing so. In the discourse on 'responsible mining', mining companies often cite generating jobs and providing amenities to communities as a demonstration of their social ethic and commitment to 'sustainability'. Doing so will simply invite scepticism. Mine closure initiatives must not be presented as part of company's (voluntary) goodwill, but should be acknowledged as what it is – compliance with regulation.

The symbolism represented by a corporation and what people have either known or experienced about the company's relations with communities, past or present, engender ingrained perceptions that become reflected in their activities. For instance, in both Acupan and Antamok, my fieldwork data, which include personal anecdotes, indicate that the proliferation of pocket mining should not be understood as a survival mechanism per se. Among many pocket miners, undertaking 'illegal' mining within company grounds is shrouded with deep-seated sentiments of attempting to somehow 'get even' or 'balance the score'. Similarly, among the small settlements surrounding Padcal, taking advantage of PMC's Social Development Management Program (SDMP) projects and attempting to make non-legitimate claims for compensation are also regarded by some individuals and certain groups as a way of making the company 'share its wealth'.

The people around mine operations, including indigenous cultural communities/indigenous peoples (ICCs/IPs) who perceive unfairness, do not passively accept loss. They too are capable of bigotry and deceit, when they perceive or experience unfairness or feel excluded. Too often, their seeming lack of knowledge is an application of know-how, including 'getting back' at those who they perceive to be 'too rich'. A mine closure program, as a development intervention, is more than 'projects of doing' (Van Ufford, Kumar Giri, & Mosse, 2003, p. 31) but a process of engaging, bridging, and accommodating viewpoints, cultures, experiences, and interests. It requires recognition that the ostensibly naïve rural folks, including ICCs/IPs, also engage in manoeuvres when they feel they need to.

Under the Mining Act, the Social Development Management Program (SDMP) is one item that local government officials, leaders of civic groups within barangays, and some families in far-flung sitios speak positively about. Equally however, some conflicts arise due to issues concerning prioritisation of specific SDMP projects over others, and allocation of projects to beneficiaries. More focus must be given to developing livelihood and income generation more than providing (physical) infrastructure projects which often become politicised, as presented in Chapter 4.

Communities are adversely affected, in various ways, by mining operations. While debates continue about the small contribution of mining to the country's nationl coffers or the export-oriented character of the industry, residents in villages have very specific concerns. They put up with insecure tenure in mining sites, drying up of water sources, damage to agricultural and residential lands, and the lack of reparation to address their plight. While international cooperative efforts are being undertaken to address the immense social, environmental and developmental issues associated with mining and mine closure, every mining company must not lose sight of local-based concerns. For example, the indigenous Ibaloy miners in Loacan want to continue their tradition-based mining for as long as they could, while migrant miners in Antamok want their illegal status to change. Meanwhile, the families in sitios around Padcal want educational scholarships, through SDMP, for their children.

The churches and NGOs have consistently provided the voice of social conscience and have tried to hold mining companies to their policies of social and environmental responsibility. As such, mine closure programs also need to recognise the role and influences of these sectors, at the same time pay attention to how communities and other social actors perceive them. The Church infuses ideas of ethics and morality into the Filipinos' religious and political culture that bring important dimensions to how policy on mineral exploitation is formulated and implemented.

Locality, belonging, and identity

The accounts of BC and PMC demonstrate that the episodes *after* mine closure (in the case of Antamok), *reopening after* closure (in the case of Acupan-Balatoc), and *anticipating* closure (in the case of Philex and the sub-villages around) bring up and accentuate issues about 'identity' and 'belonging' in relation to 'entitlement', 'rights', and 'equity.' In Philex, being 'local' as opposed to being 'migrant' becomes relevant as the local population consider themselves to be at the losing end when the 'fruits of mining' are considered. In Padcal, people talk about being Ibaloy or Kalanguya or Kankanaey compared to being a lowlander when the subject is about Padcal as the homeland of the former. Similarly, being a migrant from Mountain Province or Ifugao becomes significant to the indigenous Ibaloy in Itogon as the allocation of pocket mining sites becomes critical with the increasing numbers of 'newcomers'. In the same way, the Kalanguya in the sitios around Philex who belong to the group Indigenous People Organisation (IPO)-APSSOL have within themselves factions about authentic and non-authentic representatives, to claim royalty from PMC.

In all cases, social markers become relevant when privileges, entitlements, and rights are considered. The reason is the scarcity of resources and rewards that need allocation based on certain assumptions – whether clearly defined, accepted as given, or simply assumed. In mine closure planning, these factors must be incorporated, and initiatives pertinent to 'stakeholder management', 'community consultation' or 'community inputs' must involve dissecting different voices, values, and interests. Being recipients of certain entitlements also demand responsibility. The case of IPO-APSSOL receiving huge royalties, but failing to utilise the funds as intended, creates distrust in government policy or program. The case presents the weaknesses of the royalty system and the inexperience of the NCIP in ensuring that entitlements, as mandated by law, are matched by programs for building relevant capacities.

Large-scale corporate mines and small-scale miners

Similar to other developing countries in Africa and Latin America, the Philippines has a large-scale corporate sector, and a 'small-scale' sector. This must be recognised in any mine closure undertaking. To a large extent, the rise of the small-scale sector is a result of the decline of the large-scale sector beginning the 1980s. Historically, the traditional sites dug by inhabitants within what are now key mining districts across the country were the areas where large-scale mines began. The existence of two sectors in one location, as demonstrated by Itogon and Tuba, has important implications for considering site rehabilitation and maintenance during and after closure. Open unrehabilitated mineral lands attract other possible exploiters of a mineral resource (i.e., gold). In this situation, issues that a mining company faces is not limited to the minerals being 'stolen'. Rather, sealing off former work areas will prevent access on the grounds of public safety and health.

The accounts of pocket miners in Acupan and Antamok demonstrate the rivalry between the two sectors, with one being legal and the other illegal. While it is true that both sectors compete for the same resource, a cooperative relationship is workable. This is exemplified by the ACMP, notwithstanding its weaknesses, where both parties are willing participants in the venture. It is a negotiated arrangement for mutual coexistence.

Concluding notes

Ensuring a successful mine closure towards mine completion does not simply require unquestionable compliance to good legislation, stabilising landforms, financial guarantees, measurable and time-bound targets, among others, and well-thought strategies for monitoring and auditing. Mine closure is much larger and longer than a specific phase within the life cycle of a mine operation. It is a social juncture in the pathways and transitions, adaptations, and transformation of communities. It is a social episode during which some groups seize opportunities to partake of benefits from mining, while others bring to attention long-standing concerns, and still others assert and re-assert their identities pertinent to rights, privileges, or entitlements. The accounts of communities in Tuba and Itogon clearly suggest that mine closure cannot apply a technical-managerialist formula involving gradational or step-by-step tasks.

Notes

1 Author's interview, February 2005, Padcal, Tuba.
2 In Australia, mine closure experts P. Mulvey, A. Baker and P. Scott (2013, p.12) note that in their 30-year experience, they know of only one mine in the country that 'has been successfully closed and the lease handed back – The Mary Kathleen Uranium Mine in north-western Queensland'.

References

Australian Government Department of Industry Innovation and Science (AGDIIS). (2016, September). Mine Closure. *Leading Practice Sustainable Development Program for the Mining Industry*. Retrieved from https://industry.gov.au/resource/Documents/LPSDP/LPSDP-MineClosureCompletionHandbook.pdf

BC (Benguet Corporation). (2003). *Benguet Corporation 2002 Annual Report*. Makati City: Benguet Corporation.

BC (Benguet Corporation). (2010). The Antamok Gold Mine: A Summary of Its Remaining Potential. Retrieved from www.benguetcorp.com/index.php?option=com_content&view=article&id=181&Itemid=95

Bebbington, A. J., Bury, J. T., & Turner, B. L., II. (2009). Institutional challenges for mining and sustainability in Peru. *Proceedings of the National Academy of Sciences of the United States of America, 106*(41), 17296–17301.

Bhasin, B., & McKay, J. (2002). Mining Law and Policy in Indonesia: Reforms of the Contract of Work Model to Promote Foreign Direct Investment and Sustainability. Retrieved from http://personal.ashland.edu/~bbhasin/documents/Indonesia%20AMPLA%202002.pdf

Domingo, E. G. (2003, 8–9 October). *Public Policy Initiatives in Institutionalising CSR in the Philippine Mining Industry*. Paper presented at the World Bank International Conference on Public Policy for Corporate Social Responsibility, Washington, DC.

Griswold, D. (2000, 1 September). The Blessings and Challenges of Globalization. Retrieved from www.cato.org/publications/commentary/blessings-challenges-globalization

Haslam, P. A. (2007). The firm rules: Multinational corporation, policy space and neoliberalism. *Third World Quarterly*, *28*(6), 1167–1183.

Mines and Geosciences Bureau. (2017, January). The Philippines Minerals Industry at a Glance. Retrieved from www.mgb.gov.ph/images/homepage-images/mining-facts-and-figures-updated-January-2017.pdf

Mulvey, P., Baker, A. & Scott, P. (2012, September). Mine Closure and Waste Responsibilities and Liabilities. Retrieved from http://www.environmentalearth-sciences.com.au/wp-content/uploads/2013/01/Mine-Closure-Paper.pdf

Peck, P., Balkau, F., Bogdanovic, J., Sevaldsen, P., Skaalvik, J. F., Simonett, O., . . . & Daussa, R. (2005). Mining for Closure: Policies and Guidelines for Sustainable Mining Practice and Closure of Mines. Retrieved from http://enrin.grida.no/environment-and-security/mining_src.pdf

Ph-EITI (Philippines Environmental Tranparency Initiative). (2016, December). Extracting Value in Transarency the Third PH-EITI Report (FY2014). Retrieved from https://eiti.org/sites/default/files/documents/ph-eiti_country-report_2016_1.pdf

Presidential Decree No. 1198. (1977, 19 September). Requiring All Individuals, Partnerships or Corporations Engaged in the Exploration, Development and Exploitation of Natural Resources or in the Construction of Infrastructure Projects to Restore or Rehabilitate Areas Subject Thereof or Affected Thereby to Their Original Condition – *Full Text Published Online by the Chan Robles Virtual Law Library*. Retrieved from www.lawphil.net/statutes/presdecs/pd1977/pd_1198_1977.html

Presidential Decree No. 1251. (1977, 28 November). Imposing a Fee on Operating Mining Companies to be Known as 'Mine Wastes and Tailings Fee' to Compensate for Damages to Private Landowners and for Other Purposes. Retrieved from www.gov.ph/1977/11/28/presidential-decree-no-1251-s-1977/

van Ufford, P. Q., Kumar Giri, A. & Mosse, D. (2003). Interventions in development: Towards a new moral understanding of our experiences and agenda for the future. In P. Q. van Ufford and *A. Kumar Giri (Eds.), A Moral Critique of Development: In Search of Global Responsibilities* (pp. 1–40) London and New York: Routledge.

Index

abandoned mines 1; orphaned sites and 9; underground workings 71, 113, 121
acid rock drainage 8
Acoje 32
activists 23–24, 43
actor oriented analysis 3; approach to the mine closure process 13; intertwining relationships and interweaving domains 5
Acupan 20, 27, 80, 83–84, 127, 129–130; illegal status of pocket miners in 107–109; old and former minesite camp 100–104; pocket mining in 106–107; vista of hillocks of adits and makeshift shelters 115
Acupan-Balatoc 107, 110, 117, 125–126; current challenges in 114–117; population of 107; sub-camps and sub-villages within 103
Acupan Contract Mining Project (ACMP) 91, 101, 109–112
Acupan mine 3; closure of the 100, 105–106, 120; reopening of the 114, 124; tenement area of 104
Acupan Upper Camps Community Livelihood Association, Inc. 113
Agno River 20, 22, 42–43, 85, 104, 115
Alang 65–67, 69, 71, 75–76
alternative mining bill 28, 32, 42
Alyansa Tigil Mina 28, 30, 32
Ambalanga River 115
Ampucao 20; Barangay of 64; Ibaloy, Kankanaey and Kalanguya in 75
Anak BCI 96
ancestral lands 23, 86
Ang See, T. 18
Antamok 3, 20, 46, 54; an environmental policy borne out of 87; launchpad of industrial mining 80–83; location of 82; rich mineral lands of 83; Spanish referred to as '*Antamog*' 81
Antamokers 95
Antamok Gold Project 85–91
Antamok Miners Association 86
Apayao 55
Aquino, Benigno S. 25
Aquino, Corazon C. 22
Atlas Consolidated Mining Company 32
Atok 31, 85, 115
Australia: Lafayette project 25; Leading Practice Sustainable Development Program 7; media 9; Northern Land Council in 12
avucay 83

Bagamaspad, A. 18
Baguio: hill station 20; migration into 21, 43, 50, 52, 54, 65, 85–86, 89, 92–93, 102, 104, 106
Baguio Antamok Gold Project 85–86
Baguio Country Club 21
Baguio Gold Mining Company 43–44
Baguio Mining District 81
Baguio Water District 92
Baker, A. J. 130
baknang 83
Balatoc 30, 46, 80–82, 96, 101–102, 103, 115–116; *see also* Acupan – Balatoc

Balice, R. G. 9
balitok 81, 90
Ballard, C. 9
Balog Creek 42–43
Banks, G. 9
barangay 6, 25, 29, 46–47, 64, 66, 81, 86–87, 89–90, 117; health workers 116; officials 73, 76–77, 114, 116, 124–125, 128
battlefields of knowledge 5
Batuang 101, 103
Bauman, W. 9
belonging 11; based on locality and identity 129; close-knit relationship and strong sense of 43; identifying with an ethnolinguistic group 55; importance in claim-making 74
benefits from mining 17, 30, 57, 63, 65, 74, 122–123, 127–128; derived from company by claimant families 72; rights to and 51; tied to indigenousness 74; *see also* entitlements
Benguet Consolidated Incorporated 20
Benguet Corporation 3, 46, 80; to establish and locate an area for *Minahang Bayan* 93; history of 83–85; oldest mining company in the Philippines 46; rehabilitating the open pit 91–93; tenement area in Itogon 105
Benguet Gold Operations 3, 100; housing conditions within camps of 117; total workforce in 1988 95
Benguet Mining Tourism Village 106
Benguet Province 17; American rule in 19; former minesites in 31; PMRB of 108
best practice 12; beyond metrics of Western-scientific viewpoint 13
Bicol village 103
Bikolanos 94
Black Mountain, Inc. 32
Blaxland, D. 9
Bluestone 9
Bokod 83
Bontoc Village 103, 114
Bradbury, J. 10
Brimo, Henry A. 43–44, 48–49, 60
Buguias 64, 67
Bulawan mine 41
bunkhouses 52, 54, 58–59, 86, 94, 101–103, 105

Caballero, E. 18, 81
camote mining 83, 111
Camote Pit 93
Camote Vein 90–91
Camp Three Barangay 54–56, 64, 75; officials endorse MPSA application 76–77; part of Tuba Municipality 46; success in dispute settlement 57
cañao 49, 56, 60
Catholic Bishops Conference of the Philippines 29
Central Cordillera Forest Reserve 68
Chamber of Mines of the Philippines 22–24, 31
civil society 5–6; fact-finding mission 30; opposition to mining 28; *see also* activists
claimants 63, 68, 70; bona fide 75; illegitimate 69; for medical care 73
claims for entitlements: agreements and mechanisms in settling 69; authenticity and legitimacy of 75; criteria for assessing legitimate 70
Clarke, Metcalf A. 83
Clyde, Henry C. 83
communities: built by and evolved with mining companies 30, 32; company relations with 63; definition 6; diverse sub-groupings within 12; not homogenous 12; migrants and traditional 80; royalty to indigenous 77–78; *sitios* within mining impact zones 72; small settlements surrounding Padcal mine and Philex minesite camp 64; sub-villages and sub-camps within a Acupan-Balatoc minesite camp 103
Community Relations Division (CRD) 66; changes of staff within 75
community relations of mining company 60, 63; claims to benefits and entitlements from mining 65–66; dealing with compensation claims 66–69; multiple sub-villages around the mine 64; participation of local government officials 73–74; setting

up agreements and mechanisms in settling claims 69–73
compensation: damaged land and water 66; loss of crops 70
constitutionality: of EO 279 22; of FTAA 24; of FTAA provision and the Mining Act of 1995 3, 23–24, 122; of IPRA 23–24, 76
Constitution of the Philippines 28
Contingent Liability and Rehabilitation Fund 66
contract: expiration of mining lease 76, 85; labour 87; partnership of small-scale miners in Acupan and Benguet Corporation 109–112; secure mining 96
Cordillera: income of mine workers sent to villages 95; professionals in the 95; provinces of the 64; small-scale mining in the 32
Cordillera Administrative Region 81–82
Cordillera Peoples Alliance 51, 86, 89
culture 75; corporate 121; customs and traditions 4; transformation of rural 10

Dalicno 20, 86
Dalupirip 71
Danielson, Luke 12
Davao del Norte 108
decommissioning 2–3, 6–8, 88, 123–124; definition of 5
De Jesus, E. C. 94
DENR Administrative Order No. 1996–03 7
DENR Administrative Order No. 2015–02 88
DENR Administrative Order No. 2015–03 25
DENR Administrative Order No. 2015–07 25
DENR Administrative Order No. 2017–07 26
Department of Environment and Natural Resources 25, 27, 122; coordinated dialogues and public hearings 87; Secretary of the 22, 26, 41, 76, 93
Dick, C. 9
Digby, C. 8
DIOPIM Committee on Mining Issues 32
Dizon Copper Operation 85
Domingo E. G. 18, 31

economic dislocation 9, 48, 126
ecotourism 20, 92–93, 125
Eger, P. 12
EIA Review Committee 88, 96
engineered sanitary landfill 93
entitlements 129–130; based on locality and indigeneity 2; 'freebies' for mine workers and their families 52; indigenous people expectations and claims for 63, 65; mine employees' ranks and privileges 48, 52, 72; validity of claims for 73
environmental and social performance 1
environmental compliance certificate 42, 87
Environmental Guarantee Fund 9, 87–88
Environmental Impact Assessment 86, 88
Environmental Impact Statement 88
ethnolinguistic affiliation 94, 103; groups 32, 46, 52, 55, 114
Executive Order 79 25
Executive Order 279 22
Executive Order No. 192 27
Executive Order No. 270 23
Extractive Industries Initiative 26, 122

Fianza, Jose 83
fiesta 49, 61
Final Mine Rehabilitation and Decommissioning Plan 60
Financial and Technical Assistance Agreement 23–24
financing 7; advice of lawyer networks 13; for closure 11
Finin, G. 102
Flameygh Herman Rev. 75
Fletcher, D. 9
Fleury, A. M. 8, 11
Force, J. E. 9
free prior and informed consent 31; assertion of 76–77

Gerard, D. 9
globalisation 121

global minerals industry 1–2; investments into the Philippines 25
Global Mining Initiative 13
Gloria Macapagal Arroyo 23, 31
gold panning 27, 69–71
Gold Subsidy Law *see* Republic Act No. 6364
gold trade 18
Gozun, Elisea 41
Gumatdang 20; United Citizens of 86

Halcon, Nelia 31
Hamada-Pawid, B. 18
Harner, J. 10
Haussermann John W. 84, 101
House Bill 4315 28
House Bill No. 171 32
Howland, Harry 84
Huebert, Clinton 84
human agency 5, 125

I-Aplay 114
Ibaloy 18, 64–66, 75, 71, 114
I-Bontok 114
identity 74, 129; and increasing numbers of newcomers 129; indigenous, old settlers, former BC employees, non-employees, newcomers, sojourners 114; making distinctions between *Taga-Ditoy*, *Taga-Baba*, *Taga-Sabali* 65; and place 10; socio-cultural-spatial markers 95
Ifugao 52, 55, 64, 91, 103, 129
Igorot 72, 117; gold mines and territory 18; mine workers in Antamok 94
I-Kalinga 114
Ilocos 18, 56
Indigenous Peoples Organisation (APSSOL) 63, 69, 76–78, 129
Indigenous Peoples Rights Act 23–24, 76
indigenous persons/peoples 23, 75, 77; Ibaloy in Tuba and Itogon 20, 129; indigeneity 2, 127; indigenousness 74–76; mining royalty to 77–78
Indonesia 12, 18, 96, 121
industrial mining, beginning of 19
International Commission on Mining and Metals 7
intervention, as 'projects of doing' 5
ISO 14001 certification 25, 42, 122
Itogon: as core of the Baguio Mining District 81; location of Antamok 80; municipality of 64, 78, 88, 93, 104, 113, 116, 124, 126; a place of gold 80; population of 81; prospecting and grubstaking in 20
Itogon Inter-Barangays Alliance 87

Jackson, R. 9
jo-win 83

Kabayan 83
Kabunian 7, 90
Kalanguya 64–66, 75
Kalinga 55, 64
Kankanaey 18–20, 64, 75; specialists in small-scale mining 81
Keesing, F. M. 19
Keesing, M. 19
Kennon Road 20, 65
Keymen's Hill 103
Keystone Vein 86, 90
Kiangan village 103, 114
knowledge transfer 12
Kumar Giri, D. 5, 128

labour contractualisation 87
labour disputes 6
labour recruitment 2, 4–5, 52, 125
La Bugal B'laan 31
land: ambiguous classification of use 68; claims to 2; compensation for 68; documentation incongruous with indigenous tenure systems 20; registration introduced by the Americans 19
landform designs 8
landscapes 96, 115; altered 8; destruction of 7; economic and social 126; socio-political landscape of mining 17–32
land use 1; mine completion and alternative 123; mining as a temporary 126
Laurence, D. C. 6
leading practice *see* best practice
Legal Rights Center – Kasama sa Kalikasan 31
Leonen, Marvic 19, 31

Lepanto 30; a bronze-copper culture developed in 18; developed through LCMCo operation 46; a traditional mining locality 31
Lepanto Consolidated Mining Company 20, 46, 60
Liang River 94
life-of-mine 2, 7–8
lifeworlds, intersecting 4
livelihood 3, 63, 122; compensation for disruption of traditional 63; contract mining as alternative 110–111, 114; focus must be given to developing 128; key concerns in Barangay Virac include 116; *Minahang Bayan* where pocket miners can continue their 93; PMRB is considerate for reasons pertaining to 115; small projects to diversify 72; small-scale/pocket mining as livelihood 80, 86, 88–89; stopped due to flow of mine tailings 24
Loacan Barangay 81, 86, 89; Ibaloy miners in 93
Loacan Itogon Pocket Miners Association 86–87, 90–91, 93
Loacan Landowners Association 86; MOA with Benguet Corporation and DENR 87
Lobo, Kibungan 31–32
local government 4, 46, 92–93, 104, 111, 116, 123, 125, 129; FPIC and endorsement of 76–77; as major social actors 28–29; mining company coordination with 3, 124; mining moratoriums 42; officials' tenure of office 74; participation in SDMP 73; permitting for small-scale mining 107; revenues between national government and 122; two tiers of 66
Local Government Code *see* Republic Act No. 7160
local historic knowledge 12
locality 3–4, 74; jealousies among local population 65–66; and membership in a specific ethnolinguistic group 32; rights and entitlements tied to 2; *see also* identity
local ordinances 4, 29, 42, 116
Long N. 5, 12–13
Lopez, Gina 26, 93
Lopez, S. P. 18–19
Lynch, M. 19

Macapagal-Arroyo, G. 23
Machlis, G. E. 9
Madebwe, C. 10
Magsaysay, Ramon 68
maki-sagaok 83
Mankayan 19; many Kankanaey families left 20
mansin-eng 83
Marcopper mine 24, 28, 30, 97
Marcos, Ferdinand 21–22, 84–85, 117, 121
Marcos, Imelda 84
Masinloc Chromite Operation 85
Melchert, G. 12
memorandum of agreement 67; between company, indigenous communities, and NCIP 77; between company, small-scale miners, landowners, and DENR 87, 90
Minahang Bayan 26, 32, 93
mine closure: compliance with regulations 5; culminates in tenement relinquishment 7; definition 6; as development intervention 5; involves decommissioning and rehabilitation 5–6; literature, ideals and realities 7–13; major social actors 26–31; part of life-of-mine 8; practitioners 12; a process and social episode 2, 5; social and economic aspects of 9–10; social episode in pathways and transitions of communities 2, 5, 130; some causes of 6; support systems for 8–9
mine closure planning 1–2; approach and aim of 8; as development intervention 5; global minerals industry's policies and approaches 1; an inevitable end 6; integration of social and environmental elements in 11; social dimension in 11–12
mine completion 7–8; and alternative land use 123–125
Mineral Production Sharing Agreement 76
Mine Rehabilitation Fund Committee 66

Mines and Geosciences Bureau 17–18, 27, 42; Cordillera Administrative Region 91, 93, 110; main function and mandate of 27
minesite camp 6, 74; forerunner of modern-day 'overseas' as place to earn cash 95
Mines Minerals and Sustainable Development 12–13
mine tailings 24–25; gold from 70; intermittent leak of 94; pocket miners' containment ponds for 115; towards Balog Creek then to Agno River 42–43
mine wastes 24, 115; fees on 121; penalties for pollution from 21; pollution traced to large volumes of 8
mine workers 5–6, 9, 40–41, 94–95; 'high-grading' 107; retirement of 104; three to four generations of 103
mining: benefits and entitlements from 65; company – local government interaction 4; industrial 19
Mining Act of 1995 1, 3, 17, 22–24, 28, 97, 121–122; CBCP calls for repeal of the 29–30; civil society groups oppose 28; EIA law and 88; entitlement from mining under the 65; implementing rules and regulations of the 77; repeal of 29–30, 42, 121; SDMP under the 72
mining industry: inauspicious reputation 13; losing status as preferred land use and economic activity 1
mining in the Philippines: beginning of industrial 19; main commodities and exports 18; pre-colonial era and Spanish regime 18–19; socio-political landscape of 17–32; total tenement area of 18
Mitchell, P. 12
moratorium on mining 25, 29, 42, 92
Morga Antonio de 18, 31
Mountain Province 64, 78, 91, 103, 114, 129
Multi-partite Monitoring Team 66

National Commission on Indigenous Peoples 76–78
nationalism, local 89
NCIP Administrative Order No.2002–3 77
Negros Oriental 108
Neil, C.10
nickel 6, 8, 18, 26
no mining zone 113
non-government organisations (NGOs) 23, 27–28, 32, 86, 89, 125, 128
Noronha, L. 9
Nueva Vizcaya 67

occupational community 40, 46, 51, 74; decongestion in the 41; kith and kin 52; *see also* minesite camp
Ok Tedi 9
Ongpin, J. V. 85
open pit: in Antamok 28, 80, 86; and land compensation issues 44; operations in 'Keystone' and '440' veins: of Philex Mining Corporation 67; of Santo Niño mine 32
overseas Filipino workers 94

Pacific ring of fire 18
Padcal mine 3, 22; 20 *sitios* around 66
Pampangos 94
Pangasinense 103
Papua New Guinea 9
Paracale Gold Operation 85
Parshley, J. V. 9
Parsons, A. S. 11
Peck, P. 7. 9
peoples' organisation (POs) 23, 27–28, 63; *see also* Indigenous Peoples Organisation – APSSOL
Peterson, Nelson 83
Phelan, J. L. 18
Philex 45–47; administering community affairs 54–57; a community of kith and kin 51–52; departures of residents 57–60; divisions across ranks and privileges 52–54; population 46; residents' favour and bounty 48–51
Philex Mining Corporation 3, 40; camp decongestion 41; exploration strategy 41; extension of corporate life 41; history of 43–45; ore tonnage in 2011 41; redefined company vision 42; relations with communities 63; resumed operations after suspension

43; tailings spill of 30; tailings storage facility of 25; tenement location of 47
Philex Purok Organisation 54–55
Philippine Bill of 1902 19, 28, 85, 104–105, 117
Philippines: beginning of industrial mining 19–21; location and mineral endowment 17; mining during pre-colonial era and Spanish regime 18–19; mining policies before the Mining Act of 1995 21–22; mining tenement area 18
Philippines – Extractive Industries Transparency Initiative 26, 122
planning for closure 1; before commencing a mine 2; part of life-of-mine 8
pocket miners/pocket mining 83, 86, 89–90; in Acupan 106–107; Antamok overrun by 93; company partnership with 109–112; contract miners *versus* 112–114; grouped into cooperatives or *kumpanya* 91; illegal status of 107–109; *see also* small-scale mining
Pollet, E. A. 11
post-closure processes 7
post-mining future 2–4, 120; communities attempt to chart 9; economic development for communities 21; environmental problems associated with post-mine operation 8; offer insights into directions towards 10; post-mining landscapes 11
power relations 13; and decision-making processes 125
Presidential Decree No. 463 21, 85, 96
Presidential Decree No. 984 21, 86
Presidential Decree No. 1198 121
Presidential Decree No.1214 85, 96, 104
Presidential Decree No. 1251 121
Presidential Decree No. 1586 86, 96, 104
Presidential Decree No. 1899 107
pressure from society 1
Prill-Brett, J. 19
Pring, G. R. 1
Provincial Mining Regulatory Board 108
provincial organisations 55–56
Public Land Act of 1902 *see* Philippine Bill of 1902

Ramos, Fidel V. 22–23
ranks and privileges 52–54; and housing facilities 101–102; managers and supervisors as higher-ranking 60; and roles 43, 48
Rapu-Rapu Polymetallic Project 24; community development projects 25; tailings spill 30
Regalian Doctrine 19, 31
rehabilitation 1–3, 5–9, 31, 42–43, 80, 85, 88; of the Antamok open pit 91–93; no real commitment under old laws 121; progressive 124; 'rehabilitating' a mine area 126; restoring degraded landscapes and ecosystems 11
relinquishment 18; of a former mine area 123; mandatory 124; mine closure culminates in tenement 7
Republic Act No. 6364 21
Republic Act No. 7076 96, 108
Republic Act No. 7160 29, 81
Republic Act No. 7941 28
Republic Act No. 7942 *see* Mining Act of 1995
Republic Act No. 8731 *see* Indigenous Peoples Rights Act
retrenchment 46, 55, 58; manpower reduction 105, 107; operational restructuring program 95; voluntary retirement and 57
Rhomberg, C. 6
ritual practices 90
Romualdez, Benjamin 84
rootedness in homelands 2
Ross, N. 11
Royal Treasury of the Spanish 19
royalty 63, 129; to indigenous communities 77–78

safety and health 1, 6, 55
Sangilo 30, 32, 85
San Roque Multi-Purpose Dam 104
Santo Nino mine 31–32
Sassoon, M. 8
Scott, W. H. 18–19
Second World War 43, 54, 65, 81, 83–84, 103
Senate Bill No. 1069 28; *see also* alternative mining bill
severance payments 57, 104, 126

Silangan 41
sitio 6, 64; around Padcal mine and Philex 70
small-scale mining 29; as defined by RA No. 7076 96; displaced employees shifted to 81; not homogenous 32; subsumes other activities 83; thriving as large-scale mine operated 81–83
Smith, H. D. 12
social actors 9, 17; communities and other 128; contradicting interests of diverse 3; 'creating room for manoeuvre' 5; lifeworlds of 13; major 26–31; working with other 12
social and environmental: elements in mine closure planning 11; responsibility 64, 128; risks 8
social change 9, 117
Social Development Management Plan (SDMP) 64–65; compliance 76; education scholarships under 72; infrastructure projects 74; preferred projects of local officials 73; as required by the Mining Act 72
social disparities and conflicts 5
Sociedad anonima 83
Sociedad Minero Metalurgica Cantabro-Filipino 20
Spanish-American War 19
Surigao del Norte 41, 108

Tagalogs 94
tailings *see* mine tailings
tailings ponds 68; Dam Three 68–69, 71
Temporary No Mining Zone 113
tenure, resource access 2; concepts of property and 20; over surface land and minerals underneath 126; and patented claims 80; of pocket miners 115
TIHCAPOSI 65; negotiation with Philex Mining Corporation 72, 74
Tisza River Basin 9
Toledo City 30, 32
tradition, customs and 4; history and 125
Tuba, municipality of 46, 64, 78, 124, 130
Tublay 115
Tuding 20
Tykkyläinen, M. 10

Ucab 20, 87; mining sites in 86
Ufford, P. Q. van 5
United Highland Association 55
United Nations Environment Program 7

Virac Barangay 81, 100, 103–104, 113, 115; Council 111, 116; Ibaloy of 114; *Kapitan* of 112, 117
Visayans 94; Bisaya 103; Visayas 56, 114

Wagner, J. 12
Warhurst, A. 9, 12
water 7; bulk 92–93, 113; compensation for lost 66–67; depletion of 69; drying up of sources 128; electricity and 102, 127; farms and sources of 86; has political, economic, cultural, and spiritual dimensions 8; observed reduction and loss of 67; rights within mining claims 21; in running ball mills and sluicing gold 115; shortage of 116
watershed 104–105; reservation 108, 123
waterways: gold sediments along 83; siltation of 115
Western-scientific viewpoint 12
Wireman, M. 8
workers' mobility 2–4
workers' union 6, 48–49, 105
World Bank 7, 9

Ygolotes 18

Zambales 30, 85, 114
Zuniga, J. D. 18
Zyl van, D. 8